Lecture Notes in Computer Science 16298

Founding Editors

Gerhard Goos
Juris Hartmanis

Editorial Board Members

Elisa Bertino ⓘ, *Purdue University, West Lafayette, IN, USA*
Wen Gao, *Peking University, Beijing, China*
Bernhard Steffen ⓘ, *TU Dortmund University, Dortmund, Germany*
Moti Yung ⓘ, *Columbia University, New York, NY, USA*

The series Lecture Notes in Computer Science (LNCS), including its subseries Lecture Notes in Artificial Intelligence (LNAI) and Lecture Notes in Bioinformatics (LNBI), has established itself as a medium for the publication of new developments in computer science and information technology research, teaching, and education.

LNCS enjoys close cooperation with the computer science R & D community, the series counts many renowned academics among its volume editors and paper authors, and collaborates with prestigious societies. Its mission is to serve this international community by providing an invaluable service, mainly focused on the publication of conference and workshop proceedings and postproceedings. LNCS commenced publication in 1973.

Qi Dou · Yutong Ban · Yueming Jin ·
Sophia Bano · Mathias Unberath
Editors

Collaborative Intelligence and Autonomy in Image-Guided Surgery

First International Workshop, COLAS 2025
Held in Conjunction with MICCAI 2025
Daejeon, South Korea, September 23, 2025
Proceedings

Editors
Qi Dou
The Chinese University of Hong Kong
Hong Kong, China

Yueming Jin
National University of Singapore
Singapore, Singapore

Mathias Unberath
Johns Hopkins University
Baltimore, MD, USA

Yutong Ban
Shanghai Jiao Tong University
Shanghai, China

Sophia Bano
University College London
London, UK

ISSN 0302-9743   ISSN 1611-3349 (electronic)
Lecture Notes in Computer Science
ISBN 978-3-032-09783-5   ISBN 978-3-032-09784-2 (eBook)
https://doi.org/10.1007/978-3-032-09784-2

© The Editor(s) (if applicable) and The Author(s), under exclusive license to Springer Nature Switzerland AG 2026

This work is subject to copyright. All rights are solely and exclusively licensed by the Publisher, whether the whole or part of the material is concerned, specifically the rights of translation, reprinting, reuse of illustrations, recitation, broadcasting, reproduction on microfilms or in any other physical way, and transmission or information storage and retrieval, electronic adaptation, computer software, or by similar or dissimilar methodology now known or hereafter developed.
The use of general descriptive names, registered names, trademarks, service marks, etc. in this publication does not imply, even in the absence of a specific statement, that such names are exempt from the relevant protective laws and regulations and therefore free for general use.
The publisher, the authors and the editors are safe to assume that the advice and information in this book are believed to be true and accurate at the date of publication. Neither the publisher nor the authors or the editors give a warranty, expressed or implied, with respect to the material contained herein or for any errors or omissions that may have been made. The publisher remains neutral with regard to jurisdictional claims in published maps and institutional affiliations.

This Springer imprint is published by the registered company Springer Nature Switzerland AG
The registered company address is: Gewerbestrasse 11, 6330 Cham, Switzerland

If disposing of this product, please recycle the paper.

Preface

Artificial intelligence and robot-assisted automation technologies are rapidly transforming the landscape of various domains, including healthcare. Despite the increasing awareness of applications of the latest technologies in surgery and minimally invasive procedures, examples of domain-specific successful stories in healthcare still lag behind compared with other general fields such as autonomous driving.

The First MICCAI Workshop on Collaborative Intelligence and Autonomy in Image-Guided Surgery (COLAS 2025) provided a new venue in the MICCAI community to explore and discuss this new topic, and to foster the emerging research area. The workshop included classic CAI topics such as image-guided interventions, virtual/augmented reality, surgical planning, surgical workflow analysis, pre-/intra-operative image registration, surgical scene understanding, etc. In addition, we also covered emerging topics at the intersection of machine learning and robotics for their application in surgery, such as surgical embodied intelligence, human-robot collaborative control in medical robotics, AI-assisted autonomy in image-guided surgery, etc. Open questions such as the role of intelligent surgical assistants in next-generation operating rooms and the ethical implications of autonomous surgical systems were actively discussed. This workshop brought together researchers, surgeons, and industry leaders to discuss the latest innovations and applications of the new technologies in enhancing surgical precision, skill learning, decision-making, and patient outcomes for surgery.

At the COLAS workshop, we received strong interest from the international community, with submissions covering diverse aspects. In total, 31 submissions were carefully reviewed, resulting in the acceptance of 19 full research papers for presentation, following a double-blind peer review process in which submissions received on average three reviews each. Final decisions regarding acceptance were made by the organizing committee based on comprehensive reviewer feedback, with particular attention to scientific rigor, novelty, and relevance to the workshop themes. The selection process prioritized contributions that advanced our understanding of collaborative intelligence in surgery, whether through methodological innovations, clinical validations, or thoughtful analysis of implementation challenges.

We express our sincere gratitude to all authors for their valuable contributions and extend our appreciation to the reviewers for their dedication and constructive feedback. We also thank our keynote speakers, advisory committee members, and supporting partners for their commitment to advancing this emerging field.

September 2025

Qi Dou
Yutong Ban
Yueming Jin
Sophia Bano
Mathias Unberath

Organization

Organization Committee

Qi Dou	Chinese University of Hong Kong, China
Yutong Ban	Shanghai Jiao Tong University, China
Yueming Jin	National University of Singapore, Singapore
Sophia Bano	University College London, UK
Mathias Unberath	Johns Hopkins University, USA

Advisory Committee

Francis Creighton	Johns Hopkins University, USA
Andrew J. Hung	Cedars-Sinai Medical Center, USA
Masaru Ishii	Johns Hopkins University, USA
Nassir Navab	Technical University of Munich, Germany
Nicolas Padoy	University of Strasbourg, France
Stefanie Speidel	National Center for Tumor Diseases Dresden, Germany
Danail Stoyanov	University College London, UK
Hon Chi Yip	Chinese University of Hong Kong, China
S. Kevin Zhou	University of Science and Technology of China, China

Industry Committee

Can Zhao	Nvidia, USA
Yaokun Zhang	Intuitive Fosun, China
Kevin Koh	Vivo Surgical, Singapore

Executive Committee

Yonghao Long	Chinese University of Hong Kong, China
Chengkun Li	Chinese University of Hong Kong, China
Cheng Yuan	Shanghai Jiao Tong University, China

Contents

Towards MR-Based Trochleoplasty Planning 1
   Michael Wehrli, Alicia Durrer, Paul Friedrich, Sidaty El Hadramy,
   Edwin Li, Luana Brahaj, Carol C. Hasler, and Philippe C. Cattin

Surgical Key Step Recognition with Global-Local Modeling Mamba
in Laparoscopic Pulmonary Lobectomy 11
   Fengyue Guo, Chengkun Li, Bin Peng, Yonghao Long, Jialun Pei,
   Mengya Xu, Ziling He, Guangsuo Wang, and Qi Dou

Towards Robust Surgical Automation via Digital Twin Representations
from Foundation Models 21
   Hao Ding, Lalithkumar Seenivasan, Hongchao Shu, Grayson Byrd,
   Han Zhang, Pu Xiao, Juan Antonio Barrag, Russell H. Taylor,
   Peter Kazanzides, and Mathias Unberath

Semantic Scene Editing for Cholecystectomy Surgery 32
   Çağhan Köksal, Yousef Yeganeh, Nassir Navab, and Azade Farshad

A Training-Free Approach for 3D Reconstruction from Monocular Sinus
Endoscopy 42
   Jan Emily Mangulabnan, Roger D. Soberanis-Mukul,
   Lalithkumar Seenivasan, S. Swaroop Vedula, Masaru Ishii,
   Gregory Hager, Russell H. Taylor, and Mathias Unberath

TrackOR: Towards Personalized Intelligent Operating Rooms Through
Robust Tracking 53
   Tony Danjun Wang, Christian Heiliger, Nassir Navab,
   and Lennart Bastian

CryoFlow: Prediction of Frozen Region Growth in Kidney Cryoablation
Using a 3D Flow-Matching 64
   Siyeop Yoon, Yujin Oh, Matthew Tivnan, Sifan Song, Pengfei Jin,
   Sekeun Kim, Dufan Wu, Hyun Jin Cho, Raul Uppot, and Quanzheng Li

Temporally Stable Monocular Depth Estimation in Surgical Vision 74
   Jialang Xu, Emanuele Colleoni, Nicolas Toussaint, Muhammad Asad,
   Ricardo Sanchez-Matilla, Evangelos B. Mazomenos, Imanol Luengo,
   and Danail Stoyanov

Real-Time Surgical Keypoint Detection in Laparoscopic Cholecystectomy 85
*Yiyang You, De Ru Tsai, Yoseph Kim, Antony Goldenberg,
Juo Tung Chen, Ji Woong Brian Kim, Axel Krieger,
and Richard Jaepyeong Cha*

When Tracking Fails: Analyzing Failure Modes of SAM2 for Point-Based
Tracking in Surgical Videos 95
*Woowon Jang, Jiwon Im, Juseung Choi, Niki Rashidian,
Wesley De Neve, and Utku Ozbulak*

Deep Biomechanically-Guided Interpolation for Keypoint-Based Brain
Shift Registration 105
*Tiago Assis, Ines P. Machado, Benjamin Zwick, Nuno C. Garcia,
and Reuben Dorent*

Nested ResNet: A Vision-Based Method for Detecting the Sensing Area
of a Drop-In Gamma Probe 116
Songyu Xu, Yicheng Hu, Jionglong Su, Daniel S. Elson, and Baoru Huang

Video Grounded Conversation Generation for Reference Surgical
Instrument Segmentation 126
*Yihan Wang, Qiao Yan, Lihao Liu, Yuchen Yuan, Xiaowei Hu,
Jinpeng Li, and Pheng-Ann Heng*

Multi-stage CNN for Fast Registration of 3D Preoperative CTs to 2D
Intraoperative X-Rays 137
*Federica Facente, Benjamin Billot, Vivek Gopalakrishnan,
Manasi Kattel, Wen Wei, Polina Golland, Hervé Delingette,
Nicholas Ayache, and Pierre Berthet-Rayne*

X-RAFT: Cross-Modal Non-rigid Registration of Blue and White Light
Neurosurgical Hyperspectral Images 148
*Charlie Budd, Silvère Ségaud, Matthew Elliot, Graeme Stasiuk,
Yijing Xie, Jonathan Shapey, and Tom Vercauteren*

Cardio-Respiratory Motion Estimation and Coronary Artery Segmentation
for Image-Guided Percutaneous Coronary Intervention 158
D. China, G. Kim, N. Iyer, R. McGovern, A. Uneri, and J. Lee

Arachnoid Membrane Segmentation in Intraoperative Microscopic MVD
Surgery Scenes 168
*Jinhee Lee, Hwanhee Lee, Jay J. Park, Jeong Woo Ahn, Jong Yun Kwon,
Ciara McMahon, Julia Lewandowski, Seohee Park, Sanghoon Lee,
and Vivek P. Buch*

DARIL: When Imitation Learning Outperforms Reinforcement Learning
in Surgical Action Planning 178
   Maxence Boels, Harry Robertshaw, Thomas C. Booth,
   Prokar Dasgupta, Alejandro Granados, and Sebastien Ourselin

Temporal Propagation of Asymmetric Feature Pyramid for Surgical Scene
Segmentation 187
   Cheng Yuan and Yutong Ban

Author Index 197

Towards MR-Based Trochleoplasty Planning

Michael Wehrli[1]((✉))[iD], Alicia Durrer[1][iD], Paul Friedrich[1][iD], Sidaty El Hadramy[1][iD], Edwin Li[2][iD], Luana Brahaj[1], Carol C. Hasler[3], and Philippe C. Cattin[1][iD]

[1] Department of Biomedical Engineering, University of Basel, Allschwil, Switzerland
michaeljan.wehrli@unibas.ch
[2] Olten Cantonal Hospital, Olten, Switzerland
[3] Department of Orthopedics, University Children's Hospital Basel, Basel, Switzerland

Abstract. To treat Trochlear Dysplasia (TD), current approaches rely mainly on low-resolution clinical Magnetic Resonance (MR) scans and surgical intuition. The surgeries are planned based on surgeons experience, have limited adoption of minimally invasive techniques, and lead to inconsistent outcomes. We propose a pipeline that generates super-resolved, patient-specific 3D pseudo-healthy target morphologies from conventional clinical MR scans. First, we compute an isotropic super-resolved MR volume using an Implicit Neural Representation (INR). Next, we segment femur, tibia, patella, and fibula with a multi-label custom-trained network. Finally, we train a Wavelet Diffusion Model (WDM) to generate pseudo-healthy target morphologies of the trochlear region. In contrast to prior work producing pseudo-healthy low-resolution 3D MR images, our approach enables the generation of sub-millimeter resolved 3D shapes compatible for pre- and intraoperative use. These can serve as preoperative blueprints for reshaping the femoral groove while preserving the native patella articulation. Furthermore, and in contrast to other work, we do not require a CT for our pipeline - reducing the amount of radiation. We evaluated our approach on 25 TD patients and could show that our target morphologies significantly improve the sulcus angle (SA) and trochlear groove depth (TGD). The code and interactive visualization are available at https://wehrlimi.github.io/sr-3d-planning/.

Keywords: Implicit Neural Representation · Trochlear Dysplasia · Wavelet Diffusion Model

1 Introduction

Trochlear Dysplasia (TD) is an anatomical deformity of the femoral trochlea, often seen in adolescents with anterior knee pain and patellar instability [2]. The abnormal trochlear shape compromises patellar tracking, increasing the risk of dislocation and long-term degenerative changes, such as osteoarthritis [15]. While trochleoplasty is a recognized surgical procedure for reshaping the dysplastic trochlear groove in individuals with patellar instability, numerous patients still experience ongoing pain and exhibit varying levels of satisfaction [5]. This is largely due to the procedure's complexity, mainly qualitative planning and limited amount of discrete measurements to describe TD [24]. In current practice, TD diagnosis and surgical planning are based on MR Imaging in three orthogonal planes (axial, sagittal, coronal), each with limited resolution and spacing [22], see Fig. 1. Radiologists manually assess the sulcus angle and classify TD using the Déjour criterion [2,3]. However, intraoperative decisions are guided by the surgeon's clinical experience rather than accurate, patient-specific preoperative planning. The surgeon must reshape the femur to fit the individual's patella, a task complicated by the anatomical variability. Without consistent planning to evaluate the deformity and guide surgical decisions, achieving optimal outcomes and reproducibility becomes significantly more challenging and only experts can perform arthroscopic trochleoplasty [4].

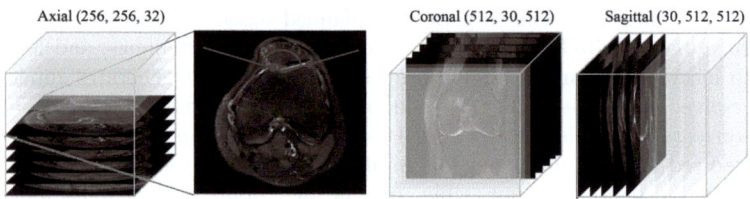

Fig. 1. Currently, the diagnosis of Trochlear Dysplasia (TD) relies on three orthogonal MR volumes: axial, sagittal, and coronal. TD shows an abnormally large sulcus angle (visualized in red) that can be seen in the axial volume. Limited resolution and inter-slice spacing impair accurate measurements. (Color figure online)

A useful preoperative plan should be based on 3D high-resolution images that allow accurate measurements and the definition of a surgical target morphology for intraoperative navigation [11]. Multiple studies have attempted to generate preoperative guidance for TD [6,7,18,24]. But all of these are based on CT scans that expose patients to harmful radiation. As trochleoplasty is often performed in adolescents whose anatomy is not yet fully developed, high-resolution, low-risk imaging is essential. A recent study [25] investigated the generation of pseudo-healthy images from MR scans using wavelet diffusion models. While this approach avoids radiation, it is limited to inpainting axial volumes, shown in the left part of Fig. 1. It can be used to give the surgeon an idea of the pseudo-healthy shape. However, it cannot support intraoperative planning, as the generated images still require segmentation, a step that is unreliable on data that has synthetic inpaintings. Segmenting synthetic images remains problematic, as they often contain out of distribution features. Popular segmentation models [8,14], trained on real data, fail to generalize to these. To overcome these limitations, we propose a novel MR-only pipeline, visualized in Fig. 2. We generate a patient-specific high-resolution 3D multi-label segmentation that reflects the expected healthy anatomy and can directly support intraoperative guidance. Our contributions are:

- A super-resolution approach leveraging implicit neural representations (INRs) to create isotropic, high-detail knee MR volumes from conventional clinical knee MR scans.
- A segment-first pipeline for pseudo-healthy 3D planning, avoiding segmentation of artificially inpainted images.
- To the best of our knowledge, our method is the first to generate surgical target morphologies for TD from MR Imaging alone. It bridges the diagnostic and surgical planning gap, promoting safer, more consistent, and potentially less invasive trochleoplasty.

2 Methods

We propose a novel MR-only workflow for the generation of patient-specific 3D target morphologies in TD, illustrated in Fig. 2. Our method proceeds in three key steps:

1. We reconstruct isotropic super-resolved 3D MR volumes from clinically acquired scans using INRs.
2. We manually annotated 40 super-resolved volumes. We trained a segmentation model with those labels to extract key anatomical structures.
3. We apply a Wavelet Diffusion Model (WDM) to inpaint the pathological region, generating a pseudo-healthy 3D volume.

The resulting inpainted multi-label segmentations can be converted into 3D meshes for intraoperative navigation. Each step is described in more detail below.

2.1 Step 1: Super Resolution MR Imaging (INR)

Clinical knee MR images are often anisotropic due to significant inter-slice spacing resulting from acquisition along different axes with high in-plane and low out-of-plane resolution. Thus, our goal is to fuse different scans into a single, high-resolution, isotropic volume. Given 3 MR images acquired along different axes (as shown in Fig. 1) and potentially different contrasts, we aim to use implicit neural representations (INRs) to fuse all available information by learning a shared representation of the underlying anatomy. Building upon [20], we define a continuous function $f_\theta : \mathbb{R}^C \to \mathbb{R}^D$ that maps a spatial coordinate $c \in \mathbb{R}^C$ to the image intensity values $d \in \mathbb{R}^D$. By jointly optimizing this function to represent all 3 scans, it implicitly learns the shared anatomical features present in the different scans, resembling a unified representation. We define this function as a 5-layer MLP with WIRE activations [23], a hidden dimension of

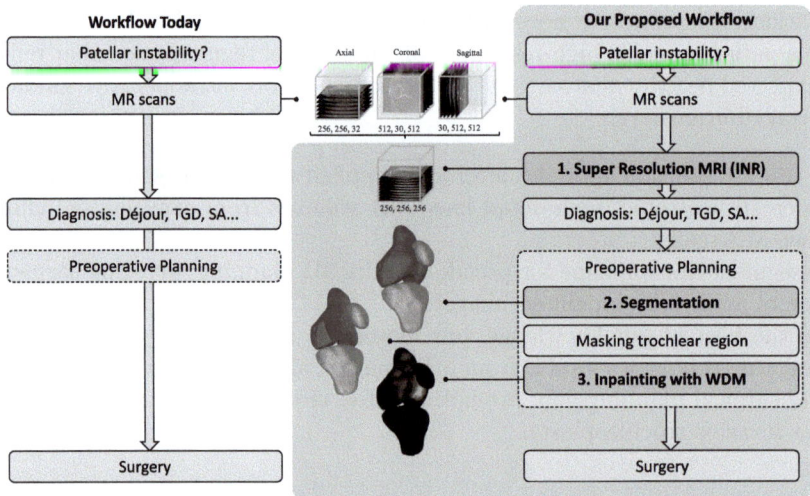

Fig. 2. Left: The workflow today. Trochleoplasty is commonly performed without preoperative planning, relying on surgical expertise. Right: Our proposed workflow enables patient-specific super-resolved 3D planning, visualized as a 3D heatmap between the patients original knee with TD and it's pseudo-healthy version.

1024, and 3 output channels (one per scan), and jointly optimize it, by minimizing the pixel-wise mean-squared error (MSE) between the model's prediction, and a given value within a set of known pixel-value pairs $\mathcal{C}$:

$$\mathcal{L}_{MSE} = \sum_{c \in \mathcal{C}} \|f_\theta(c) - d(c)\|_2^2. \tag{1}$$

Each model is optimized for 100 epochs, with a batch size of 4096 pixel-value pairs per iteration. We use an Adam optimizer with a learning rate of 4×10^{-4} that follows a cosine annealing learning rate schedule. To obtain a super-resolved volume, we evaluate the learned function over a given spatial domain, using an arbitrary resolution (in our case 256^3).

2.2 Step 2: Segmentation

We annotated 10 super-resolved MR volumes from our in-house dataset and 30 super-resolved MR volumes from the fastMRI dataset, see Sect. 3. Four anatomical structures were labeled: Femur, tibia, patella and fibula. We used 3D Slicer [1,16] with the MONAI Label framework [9]. The manual segmentations were reviewed by a deputy attending orthopedic surgeon. We chose a 3D U-Net (with a pretrained SegResNet [21]) using a Dice and cross-entropy loss.

2.3 Step 3: Inpainting with WDM

Unlike [10,25], who inpaint image intensities, we first segment the volume (as explained in Sect. 2.2) and then do inpainting in the segmented space. We argue that this simplifies the enforcement of anatomical priors, since the final objective is to obtain a target morphology. Before using the Wavelet Diffusion Model (WDM) by [13] we first mask the trochlear region (with a 30 mm offset around the patella), as shown in Fig. 2 between Step 2 and Step 3, resulting in our condition m used to train the WDM, similar to [12]. Figure 3 shows our adapted version of the WDM for inpainting: The wavelet coefficients x_0 are obtained through a DWT of the ground truth y_0. Noise is added to x_0, resulting in $x_t = \sqrt{\bar{\alpha}_t} x_0 + \sqrt{1-\bar{\alpha}_t}\epsilon$, where $\epsilon \sim \mathcal{N}(0, \mathbf{I})$ with $\alpha_t = 1 - \beta_t$ and $\bar{\alpha}_t = \prod_{s=1}^{t} \alpha_s$. Concatenating x_t and the wavelet transformed masked image m results in $X_t := (x_t, DWT(m))$. During training, t is randomly chosen at each iteration. X_t and t are the inputs to the neural network ϵ_θ, which predicts the denoised $\tilde{x}_0$. The model is trained using the MSE loss between $\tilde{x}_0$ and x_0. During inference, the model is iteratively evaluated for $t = T, ..., 1$ with $T = 1000$. Afterwards, the final prediction is obtained through an IDWT of $\tilde{x}_0$.

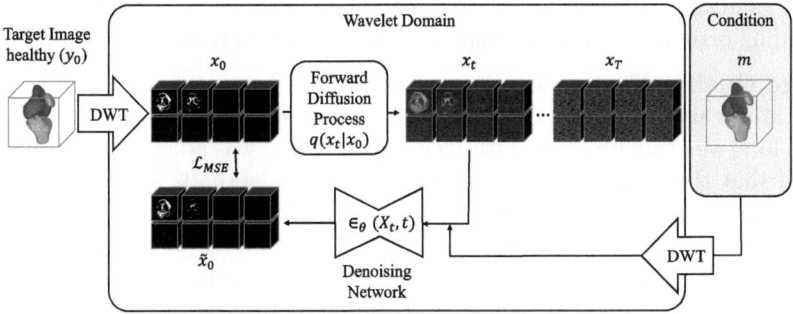

Fig. 3. During training, the network is tasked with predicting the denoised image $\tilde{x}_0$ based on $X_t := (x_t, DWT(m))$ at a random time step t. The loss is then calculated between the prediction $\tilde{x}_0$ and x_0. Figure inspired by [12].

All used images have a resolution of 256^3. The network was trained with batch size 1 and learning rate 1×10^{-5} until convergence (50'000 iterations). T was set to 1000. More implementation details can be found at https://wehrlimi.github.io/sr-3d-planning/. For visualization purposes we use Marching Cubes to convert the inpainted segmentation to a mesh [19].

3 Experiments

Data. We evaluated our pipeline using two datasets: The publicly available fastMRI dataset (https://fastmri.med.nyu.edu/) [17, 26] and an in-house dataset for patients diagnosed with TD. The entire pipeline, up to Step 3 (see Fig. 2), is applied identically to both datasets. For Step 3, we exclusively used data from fastMRI to train. We considered only subjects that have an axial proton density fat-saturated sequence and have all 3 volumes. Resulting in 1'568 subjects. We assume that this dataset represents a healthy trochlear shape representation. We split it in 80% (1'254) for training, 20% for testing of the model. Additionally, we used our in-house dataset with TD patients to evaluate our workflow on pathological data. It was anonymized and granted an exemption by the Ethikkommission Nordwest- und Zentralschweiz (Req-2024-01188). The described methods were carried out according to the relevant guidelines and regulations. From the available 52 patients, we excluded 27 patients having a subluxated patella (during the MR scan) or intense swelling (20) or no registered volumes (7) - Step 1 does not work then. The trained WDM was applied for inference to 25 TD patients (26 scans) to generate pseudo-healthy plans, offering insight into how their trochlear region might appear in the absence of pathology. The final image in Fig. 2 illustrates a 3D anomaly map, clearly indicating by bright colors that the patient's trochlea is abnormally shallow.

Medical Evaluation. We compared 26 knee MR scans of 25 patients with TD before and after inpainting using sulcus angle (SA), trochlear groove depth (TGD) and Déjour classification. The measurements were performed by a deputy attending orthopedic surgeon using 3D Slicer [1,16] Version 5.6.2.

Image Quality Evaluation. Due to the lack of ground-truth high-resolution knee MR scans, we were not able to quantitatively verify Step 1 on our data. Therefore, we evaluated it qualitatively. The method was adapted from [20], who stated that they surpass State-of-the-Art (SOTA) methods in super-resolution reconstruction. To evaluate Step 2, we used the Dice Score. The super-resolved subset (see Sect. 2.2) was randomly split into 90% training (36 volumes) and 10% test (4 volumes). To evaluate Step 3, we masked out a healthy knee region in the fastMRI test set, inpainted it using our proposed approach, and measured the deviation from the, in this case known, ground truth morphology using MSE.

4 Results

4.1 Medical Evaluation

Of 26 pathological scans, 14 showed a clear reduction in the severity of TD in the pseudo-healthy versions according to the Déjour criterion by one or more stages. Figure 4 (right) demonstrates this shift. None of the samples were rated more severe than before. In 22 patients the SA and TGD could be measured before (due to the nature of TD) and after inpainting. Regarding SA, a significant shift to a less shallow trochlea (from a mean of 162° to a mean of 154°) was observed (p value = 0.0011) by the Wilcoxon signed-rank test. The increase of the TGD is significant as well (p = 0.00006, from a mean of 1.48 mm to 2.33 mm). A qualitative example is visualized in Fig. 5.

4.2 Image Quality Evaluation

Step 1: INR. Qualitative results for a patient with TD can be seen in Fig. 6. The very right column shows the INR volume, that has high-resolution images in all 3 views.

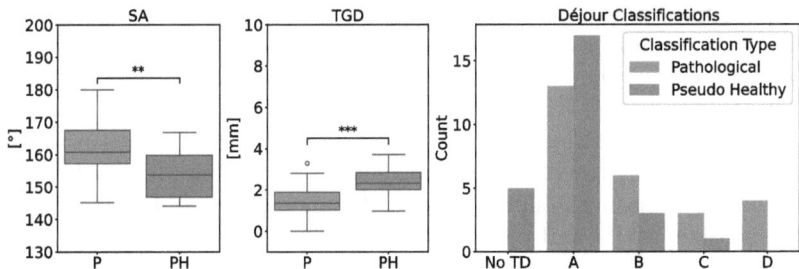

Fig. 4. Left: The difference between the SA in the 22 scans is significant. Middle: Significant difference in TGD. Right: The pseudo-healthy versions show an evident reduction in severity according to the Déjour criterion.

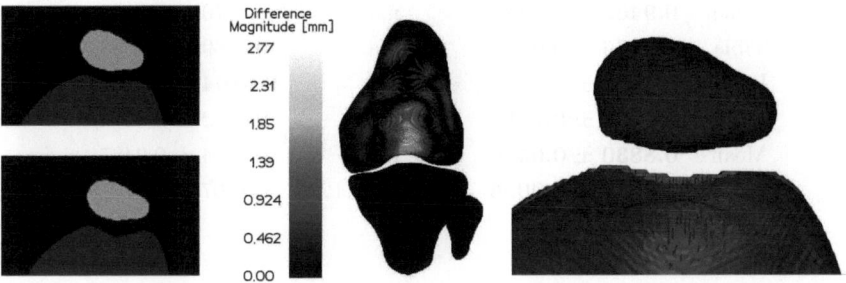

Fig. 5. Left (top): Segmentation before inpainting of a patient with TD. Left (bottom): After inpainting. Middle: Difference Magnitude in 3D of the same patient, visualized w.o. patella. Right: Axial view that shows the difference between the original and morphological target. Blue arrows indicate where bone has to be removed. Red arrows indicate where bone could be added. (Color figure online)

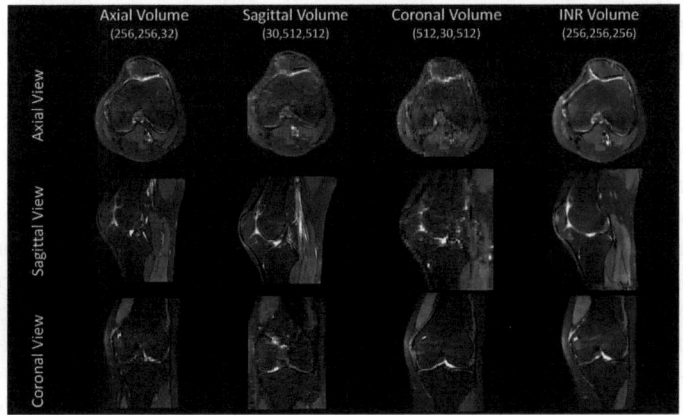

Fig. 6. Left to right: Axial volume, Sagittal volume, Coronal volume and on the very right the INR volume, combining the information of all others.

Step 2: Segmentation. We compare our custom-trained segmentation model (MonaiLabel) against SOTA bone knee segmentation tools. Note that we are not claiming to have the best performing segmentation model architecture. We demonstrate, see Table 1, that with our manual segmentations one can achieve superior femur segmentation in super-resolved knee MR scans - a crucial step to create pseudo-healthy trochleas. Totalsegmentator [8] has no module to segment the full knee MR scan at once. We used the `appendicular_bones_mr` module to segment tibia, fibula and patella and `total_mr` for the femur. SegmentAnyBone [14] is only able to perform binary segmentation.

Table 1. Quantitative segmentation results (Dice score ± std).

	MonaiLabel (Ours)	SegmentAnyBone	TotalSegmentator
Femur	**0.9402 ± 0.0187**	n.a.	0.8970 ± 0.0351
Tibia	**0.9266 ± 0.0089**	n.a.	0.8869 ± 0.0333
Patella	0.7870 ± 0.1087	n.a.	**0.8104 ± 0.0593**
Fibula	**0.8784 ± 0.0699**	n.a.	0.7232 ± 0.0672
Mean	**0.8830 ± 0.0516**	n.a.	0.8294 ± 0.0487
Binary	**0.9262 ± 0.0096**	0.7392 ± 0.1276	0.8807 ± 0.0181

Step 3: Wavelet Diffusion Model On the fastMRI test set, we observed an MSE of: 0.0017 ± 0.0015.

5 Conclusion

We propose a pipeline for the generation of surgical target morphologies for trochleoplasty from conventional clinical MR scans. The pipeline has the potential to be used intraoperatively for surgical navigation. While this work focused on TD, the pipeline is not limited to this pathology and could be adapted to other surgical applications. However, our method also comes with certain limitations: A sufficiently large dataset representing healthy anatomy is required to train the WDM. Further, the fastMRI dataset may not capture the ideal trochlear shape, could be age-biased, and is likely skewed toward a North American population. Additionally, conventional MR scans must be registered to be used. Also, scans with a dislocated patella are not suitable for our current approach. Segmentation quality remains a bottleneck, and could be improved with more extensive manual annotations. In conclusion, we propose a pipeline that generates super-resolved, patient-specific 3D pseudo-healthy target morphologies from conventional MR scans. It paves the way for more reproducible, minimally invasive and patient-tailored surgery procedures. The next major step is to validate this pipeline intraoperatively using a navigation system, moving toward integration into real-world surgical workflows.

Acknowledgments. This work was financially supported by the Werner Siemens Foundation through the MIRACLE II project.

Disclosure of Interests. The authors have no competing interests to declare that are relevant to the content of this article.

References

1. 3D Slicer Community: 3d slicer - a free, open source, and extensible image computing platform. https://www.slicer.org/ (2025), Accessed 03 Jul 2025
2. Batailler, C., Neyret, P.: Trochlear dysplasia: imaging and treatment options. EFORT Open Rev. **3**(5), 240–247 (2018)
3. Beaufils, P., Thaunat, M., Pujol, N., Scheffler, S., Rossi, R., Carmont, M.: Trochleoplasty in major trochlear dysplasia: current concepts. Sports Med. Arthroscopy Rehabil. Therapy Technol. **4**, 1–8 (2012)
4. Blønd, L.: Arthroscopic deepening trochleoplasty: the technique. Operative Tech. Sports Med. **23**(2), 136–142 (2015)
5. Blønd, L., Barfod, K.W.: Trochlear shape and patient-reported outcomes after arthroscopic deepening trochleoplasty and medial patellofemoral ligament reconstruction: a retrospective cohort study including mri assessments of the trochlear groove. Orthop. J. Sports Med. **11**(5), 23259671231171376 (2023)
6. Cerveri, P., Belfatto, A., Baroni, G., Manzotti, A.: Stacked sparse autoencoder networks and statistical shape models for automatic staging of distal femur trochlear dysplasia. Int. J. Med. Robot. Comput. Assisted Surgery **14**(6), e1947 (2018)
7. Cerveri, P., Belfatto, A., Manzotti, A.: Representative 3d shape of the distal femur, modes of variation and relationship with abnormality of the trochlear region. J. Biomech. **94**, 67–74 (2019)
8. D'Antonoli, T.A., et al.: Totalsegmentator mri: robust sequence-independent segmentation of multiple anatomic structures in mri. Radiology **314**(2), e241613 (2025)
9. Diaz-Pinto, A., Alle, S., Nath, V., Tang, Y., Ihsani, A., Asad, M., Pérez-García, F., Mehta, P., Li, W., Flores, M., et al.: Monai label: a framework for ai-assisted interactive labeling of 3d medical images. Med. Image Anal. **95**, 103207 (2024)
10. Durrer, A., et al.: Denoising diffusion models for 3d healthy brain tissue inpainting. In: MICCAI Workshop on Deep Generative Models, pp. 87–97. Springer (2024)
11. Fang, X., et al.: Patient-specific reference model estimation for orthognathic surgical planning. Int. J. Comput. Assist. Radiol. Surg. **19**(7), 1439–1447 (2024)
12. Friedrich, P., Durrer, A., Wolleb, J., Cattin, P.C.: cwdm: conditional wavelet diffusion models for cross-modality 3d medical image synthesis. arXiv preprint arXiv:2411.17203 (2024)
13. Friedrich, P., Wolleb, J., Bieder, F., Durrer, A., Cattin, P.C.: Wdm: 3d wavelet diffusion models for high-resolution medical image synthesis. In: MICCAI Workshop on Deep Generative Models, pp. 11–21. Springer (2024)
14. Gu, H., et al.: Segmentanybone: a universal model that segments any bone at any location on mri. Med. Image Anal. 103469 (2025)
15. Hasler, C.C., Studer, D.: Patella instability in children and adolescents. EFORT Open Rev. **1**(5), 160–166 (2016)
16. Kikinis, R., Pieper, S.D., Vosburgh, K.G.: 3d slicer: a platform for subject-specific image analysis, visualization, and clinical support. In: Intraoperative Imaging and Image-Guided Therapy, pp. 277–289. Springer (2013)

17. Knoll, F., et al.: fastmri: a publicly available raw k-space and dicom dataset of knee images for accelerated mr image reconstruction using machine learning. Radiol. Artif. Intell. **2**(1), e190007 (2020)
18. Lee, J.Y., et al.: Shallow trochlear groove and narrow medial trochlear width at the proximal trochlea in patients with trochlear dysplasia: a three-dimensional computed tomography analysis. Knee Surgery, Sports Traumatology, Arthroscopy **32**(6), 1434–1445 (2024)
19. Lorensen, W.E., Cline, H.E.: Marching cubes: a high resolution 3d surface construction algorithm. In: Seminal Graphics: Pioneering Efforts that Shaped the Field, pp. 347–353 (1998)
20. McGinnis, J., et al.: Single-subject multi-contrast mri super-resolution via implicit neural representations. In: International Conference on Medical Image Computing and Computer-Assisted Intervention, pp. 173–183. Springer (2023)
21. Myronenko, A.: 3d mri brain tumor segmentation using autoencoder regularization. In: International MICCAI Brainlesion Workshop, pp. 311–320. Springer (2018)
22. Nacey, N.C., Geeslin, M.G., Miller, G.W., Pierce, J.L.: Magnetic resonance imaging of the knee: an overview and update of conventional and state of the art imaging. J. Magn. Reson. Imaging **45**(5), 1257–1275 (2017)
23. Saragadam, V., LeJeune, D., Tan, J., Balakrishnan, G., Veeraraghavan, A., Baraniuk, R.G.: Wire: wavelet implicit neural representations. In: Proceedings of the IEEE/CVF Conference on Computer Vision and Pattern Recognition, pp. 18507–18516 (2023)
24. Van Haver, A., et al.: A statistical shape model of trochlear dysplasia of the knee. Knee **21**(2), 518–523 (2014)
25. Wehrli, M., et al.: Generating 3d pseudo-healthy knee mr images to support trochleoplasty planning. Int. J. Comput. Assist. Radiol. Surgery 1–8 (2025)
26. Zbontar, J., et al.: fastmri: an open dataset and benchmarks for accelerated mri. arXiv preprint arXiv:1811.08839 (2018)

Surgical Key Step Recognition with Global-Local Modeling Mamba in Laparoscopic Pulmonary Lobectomy

Fengyue Guo[1(✉)], Chengkun Li[1], Bin Peng[3], Yonghao Long[1], Jialun Pei[1], Mengya Xu[1], Ziling He[2], Guangsuo Wang[3], and Qi Dou[1]

[1] The Chinese University of Hong Kong, Hong Kong SAR, China
qidou@cuhk.edu.hk
[2] Jinan University, Guangzhou, China
[3] Department of Thoracic Surgery, Shenzhen People's Hospital, Shenzhen, China
wang.guangsuo@szhospital.com

Abstract. Accurate surgical phase recognition in laparoscopic pulmonary lobectomy is essential for workflow optimization, intraoperative guidance, and surgical training. However, high procedural variability and visual similarity between key phases (vessel dissection and bronchus dissection) pose significant challenges. To address this, we target a subset of fine-grained steps selected for their clinical relevance and distinct visual features. We propose a unified framework that integrates global temporal modeling and local feature refinement, capturing long-range procedural dependencies and fine-grained spatial dynamics. We also construct a clinically validated dataset covering diverse lobectomy subtypes and intraoperative variations with standardized annotations. By modeling temporal continuity and refining local anatomical features, our method enables robust recognition across complex surgical scenes. Experiments show that it achieves 92.15% accuracy, surpassing prior state-of-the-art methods across several major metrics. The proposed framework shows potential for supporting future research on real-time surgical workflow analysis and remote surgery assistance.

Keywords: Laparoscopic Video Analysis · State Space Model

1 Introduction

Pulmonary lobectomy is a key treatment for lung cancer and other thoracic diseases. Minimally invasive approaches, including video- and robotic-assisted thoracic surgery, are increasingly favored over open surgery due to smaller incisions, fewer complications, and faster recovery [8,11]. However, these techniques demand high technical skill due to limited workspace and complex anatomy.

Co-corresponding authors—G. Wang and Q. Dou

Accurate workflow recognition is essential for intraoperative decision making and reducing complication risks through context-aware assistance [3,7,12].

Compared to workflow recognition studies in relatively standardized surgeries like cholecystectomy, hysterectomy and cataracts [13,15,18], pulmonary lobectomy remains underexplored in the field of surgical workflow analysis. The complexity of lobectomy stems from several factors. First, the surgery involves multiple subtypes (e.g., left upper, left lower, right upper, right middle, and right lower lobectomies), each with unique anatomical and procedural variations. Second, the sequence of surgical steps may dynamically change depending on the patient's anatomy or the surgeon's preferred technique (e.g., fissure-first, anterior-to-posterior, or posterior-to-anterior) [10]. Prior works on surgical phase recognition have continuously improved temporal modeling. For instance, TMR-Net [5] introduced a non-local temporal bank, but its use of simple weighted aggregation limited global context capture. TeCNO [1] subsequently extended temporal receptive fields using dilated convolutions, though at the expense of fine-grained temporal resolution. Transformer-based models, such as Trans-SVNet [6], leveraged attention mechanisms to enhance global temporal modeling, yet processed spatial and temporal features separately, introducing redundancy. Surgformer [16] further advanced this by unifying spatial-temporal learning via hierarchical temporal structures and divided attention, enhancing adaptability to complex workflows, though it still risks overlooking critical local visual cues in procedures like lobectomy. Recently, structured state-space models, particularly the Mamba architecture [4], have emerged as efficient alternatives, capable of linear-time modeling of long sequences. For example, SPRMamba [17] integrated structured memory with scaled residual blocks to improve fine-grained feature representation in endoscopic submucosal dissection. Beyond phase-level recognition, approaches like MTMS-TCN [14] adopted hierarchical frameworks to explicitly distinguish between long-term surgical phases and short-term steps, emphasizing the helpfulness of identifying subtle, short-term steps for surgery. Nonetheless, current methods struggle with accurately recognizing these fine-grained phases due to insufficient local feature representation and limited temporal resolution. Additionally, the absence of publicly available datasets for pulmonary lobectomy further restricts targeted research addressing these challenges.

To address these challenges, we construct a dedicated dataset for pulmonary lobectomy, covering diverse subtypes and real-world cases with intraoperative conversions. Expert annotations define six key workflow steps, providing a standardized yet flexible framework for temporal analysis. Building on Mamba's efficiency in global sequence modeling, we introduce a Local Feature Refinement Block (LFRB), which applies shallow Mamba layers to spatial patch tokens to enhance fine-grained features smoothed during global aggregation. These local features are fused back into the global token stream, balancing context with critical local detail. Together, the global Mamba backbone and LFRB form a unified framework tailored to the visual and structural complexities of lobectomy key step recognition. The main contributions of this work are summarized as follows:

(1) We propose the first Mamba-based framework specifically designed for video analysis in laparoscopic pulmonary lobectomy.
(2) We adopt a global-local feature modeling approach, combining local refinement for detailed spatial understanding with global context modeling to capture procedural coherence.
(3) We construct a new dataset for surgical workflow analysis in minimally invasive pulmonary lobectomy, including both video-assisted and robotic-assisted procedures, covering diverse procedural subtypes and intraoperative variations.

2 Method

2.1 Overview

Given a lobectomy video stream, we first uniformly sample a sequence of T frames at 1 fps to construct a compact but informative video clip represented as $V \in \mathbb{R}^{3 \times T \times H \times W}$, where $C = 3$ denotes the number of color channels, T the temporal length, and (H, W) the spatial resolution per frame. This sampling strategy reduces temporal redundancy while preserving essential surgical details. Each frame is independently partitioned into non-overlapping spatial patches using a 2D convolutional projection with kernel size 16×16 and stride 16×16, shared across all frames. The resulting patches are flattened and projected into embedding vectors, forming a unified sequence of patch tokens across the temporal dimension. A learnable [CLS] token is inserted at the mid-position of each frame, and absolute spatial-temporal positional embeddings are added to retain structural information.

To capture spatial-temporal dependencies across the entire clip, we replace traditional Transformer blocks with a stack of Mamba-based sequence modeling blocks. These structured state-space layers enable efficient modeling of both short- and long-range interactions in the token sequence. An optional bidirectional variant further enhances global temporal context by aggregating information in both forward and backward directions. To improve local feature fidelity—especially for subtle or transient surgical cues—a lightweight local refinement module composed of shallow Mamba blocks processes non-[CLS] patch tokens. The resulting local features are fused back into the global token stream to enrich fine-grained representations. Finally, [CLS] tokens from each frame are aggregated and passed through a classification head to predict the step. An overview of the full framework is shown in Fig. 1.

2.2 Global Temporal Modeling

To extract global spatial-temporal features from the video, we adopt a stack of sequence modeling blocks based on the Mamba architecture. After patch embedding, each frame produces a fixed-length sequence of spatial tokens. A learnable [CLS] token is inserted at the middle position of each frame to represent its global semantics, and absolute spatial-temporal positional embeddings are added to

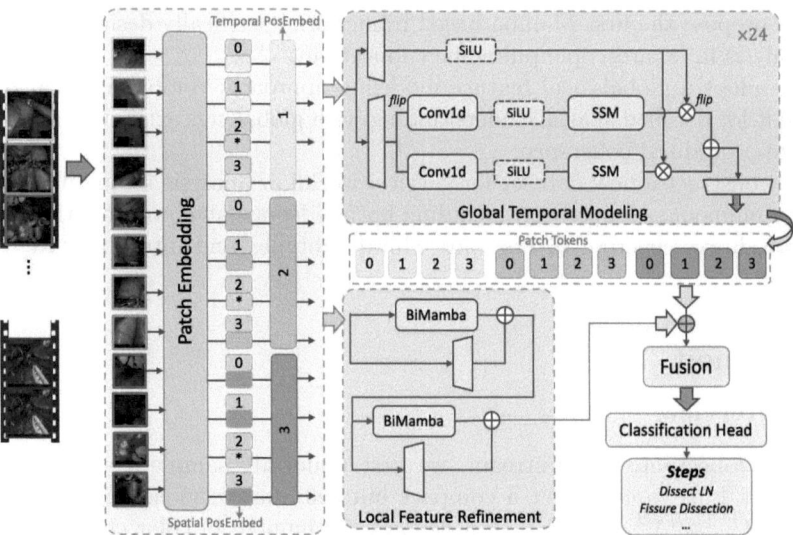

Fig. 1. The input video is first embedded into patches, which are processed by a global temporal branch. Then the same patches go through a local refinement branch. The outputs are fused and passed to the classification head for prediction.

preserve structural information. The entire sequence, composed of patch tokens and inserted [CLS] tokens across all frames, is flattened and processed jointly by a series of Mamba blocks. Each block consists of a normalization layer, a structured state-space operator, and a residual connection. Optional enhancements such as drop path and fused residual normalization are used to improve training stability and convergence. By modeling the full token sequence in a unified manner, this design facilitates spatial and temporal information flow across frames. The final [CLS] tokens are extracted and aggregated to form a global video representation, which is then passed to a classification head for the prediction of the surgical step.

2.3 Local Feature Refinement

While global sequence modeling provides a comprehensive understanding of procedural progress, it may overlook small spatial features that are crucial for accurately identifying surgical step transitions. This limitation is particularly evident in lobectomy procedures, where stages such as vessel and bronchus dissection involve anatomically adjacent structures with highly similar visual appearances. Distinguishing features mainly rely on subtle differences in tissue texture and morphology. Global modeling approaches suppress the preservation of such local cues, reducing the model's ability to accurately distinguish between key steps.

To address this, we introduce a Local Feature Refinement Block, operating exclusively on the patch tokens $X_1^{\text{patches}} \in \mathbb{R}^{L \times C}$. The Local Refinement module consists of shallow Mamba layers, designed to capture short-range spatial and temporal patterns efficiently. Given the input patch tokens, local features are extracted as:

$$X_{\text{local}} = \text{LocalRefine}(X_1^{\text{patches}}), \quad (1)$$

where X_{local} captures fine-grained motion cues and anatomical textures across neighboring patches.

To integrate global and local information, we combine X_{local} with the globally modeled patch features X_2, obtained from the Global Modeling Block , via element-wise addition:

$$X_{\text{final}} = X_2 + X_{\text{local}}, \quad (2)$$

where the updated [CLS] token from X_2 is retained for phase classification.

3 Experiment

3.1 Experimental Settings

Datasets. There is currently no publicly available dataset focused specifically on lobectomy surgery. To advance research in surgical workflow analysis, we collected a new dataset comprising steps videos of robot-assisted lobectomy (RAL) and video-assisted thoracoscopic segmentectomy (VATS-segmentectomy). Both surgical approaches are commonly employed in thoracic surgery and may involve intraoperative conversion due to anatomical complexity or dissection difficulty. The inclusion of both types increases the variety of the dataset and helps the model handle different surgical situations (Table 1).

Table 1. Annotation protocol for each surgical step

Step	Annotation Protocol
Preparation	Setup steps before procedure begins. Start: first internal scene. End: instruments placed, just before operative step
Dissect LN	Remove lymph nodes from stations 7–12 (2,4 on right; 5,6 on left). Start: first contact with node tissue. End: node fully removed
Fissure Dissection	Separate lung lobes using cautery or stapler. Start: first gesture orienting fissure. End: fissure completely divided
Vessel Dissection	Expose and divide vessels supplying affected lobe using stapler. Start: first interaction with vessel area. End: stapler activated
Bronchus Dissection	Isolate and divide bronchus using stapler. Start: first contact with bronchus tissue. End: stapler opened
Specimen Removal	Remove tissue with retrieval bag. Start: bag placement. End: bag fully removed from body

We collected a proprietary dataset of 361 surgical video clips from a collaborating hospital and internal research contributors. All videos were de-identified

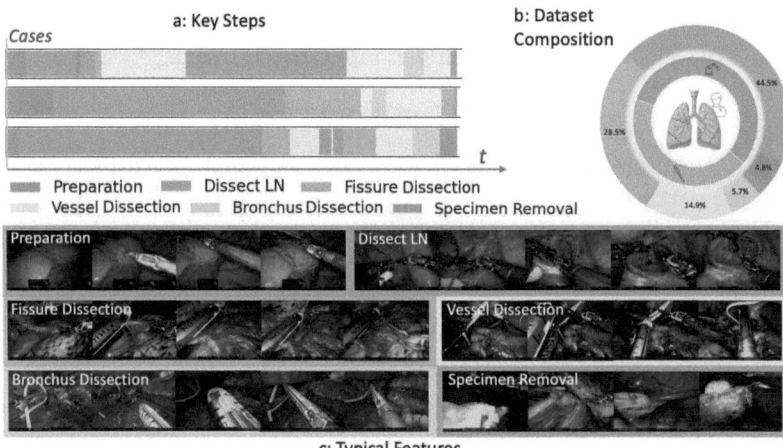

Fig. 2. Visualization of key steps annotations and dataset structure. a Examples from three robotic-assisted lobectomy videos. b Dataset composition, with the inner pie chart showing robotic-assisted (blue) vs. thoracoscopic (orange) case proportions. c Representative features per steps; workflow flows left to right, top to bottom. Steps colors are consistent and follow the central legend. (Color figure online)

and selected under the supervision of experienced thoracic surgeons. The dataset includes 201 robotic and 160 endoscopic cases, encompassing both typical and complex intraoperative scenarios. In total, over 60,312 frames were recorded at 30 fps using high-definition surgical cameras, with robotic procedures captured at 720 × 480 pixels resolution and endoscopic procedures at 1920 × 1080 pixels. All videos were stored in MPG/MP4 format and fully anonymized to ensure.

The lobectomy videos are annotated and divided into six key surgical steps: (1) Preparation; (2) Dissect LN; (3) Fissure Dissection; (4) Vessel Dissection; (5) Bronchus Dissection; (6) Specimen Removal. These annotations cover critical intraoperative stages and take into account both procedural consistency and variations arising from intraoperative conversions. The composition of the dataset and the characteristics of different stages are shown in Fig. 2. The data set is divided into 65 training sets, 5 validation sets, and 10 test sets. All data sets frames resized to 250 × 250 pixels for preprocessing.

Evaluation Metric. Following previous works on phase recognition [1,2,16], we used four standard evaluation metrics to quantitatively assess model performance: Accuracy, Precision, Recall and Jaccard Index. Furthermore, we include the F1-score in the ablation study, which offers a balanced measure of both precision and recall.

Implementation Details. The model is trained for 45 epochs on a single NVIDIA GeForce RTX 3090 GPU with a batch size of 4. Shared weights are initialized from a Kinetics-400 pre-trained model [19], while the remaining parameters are randomly initialized. Training is performed using the AdamW optimizer with momentum coefficients $\beta_1 = 0.9$ and $\beta_2 = 0.999$, an initial learning rate of 2×10^{-4}, and a layer-wise decay rate of 0.8. Each input clip consists of 16 consecutive frames sampled at a temporal stride of $R = 1$. Mixed-precision training with BFloat16 is employed to reduce memory usage and accelerate computation. To address class imbalance, we apply a custom class-weighted cross-entropy loss. Class weights are computed based on inverse frequency, clamped to a maximum value, smoothed with a $\log(1 + x)$ transformation, and normalized. The weight of 'Bronchus Dissection' is further increased due to its difficulty in recognition. During training, weighted soft target cross-entropy is used with mixup augmentation, while weighted label smoothing cross-entropy is applied otherwise.

3.2 Comparison with State-of-the-Art Methods

We evaluate the performance of our framework against several state-of-the-art approaches for surgical workflow recognition, including TeCNO [1], TransSVNet [6], Surgformer [16] and VideoMamba [9]. All baseline models are implemented using their official repositories to ensure fair comparisons under consistent settings.

As shown in Table 2, our framework achieves state-of-the-art performance on the robotic lobectomy dataset, with an accuracy of 92.15%, outperforming all compared methods. It surpasses the second-best model, VideoMamba, by 2.49% in accuracy and shows consistent gains across precision (+2.40%), recall (+2.48%), F1-score (+2.85%), and Jaccard index (+4.31%). Figure 3 visualizes the learned feature distributions, where our framework yields more compact and distinct clusters, particularly improving separation between commonly confused steps such as Bronchus and Vessel Dissection. The macro-average precision-recall curve further indicates higher precision at moderate-to-high recall levels, despite a comparable average precision (0.83) to VideoMamba. Finally, Fig. 3c presents sstep recognition results on two representative robotic lobectomy videos.

Table 2. Performance comparison with the previous state-of-the-art methods for workflow recognition and video recognition.

Model	Accuracy(%)	Precision(%)	Recall(%)	F1-score(%)	Jaccard(%)
TeCNO [1]	79.30	84.17	79.85	78.53	47.35
TransSVNet [2]	84.07	83.87	84.07	82.52	73.98
Surgformer [16]	88.37	87.34	88.37	86.69	81.06
VideoMamba [9]	89.66	88.48	89.66	88.51	82.67
Ours	**92.15**	**90.88**	**92.14**	**91.36**	**86.98**

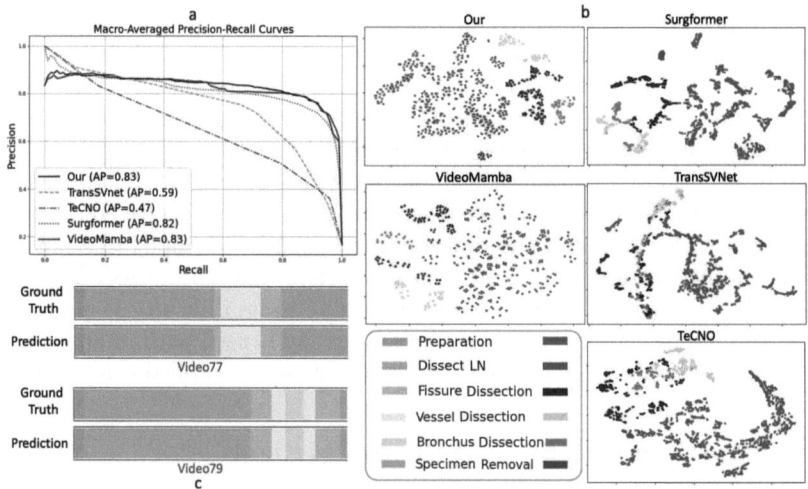

Fig. 3. a: The Macro-Averaged Precision-Recall Curves of five methods. b: Visualization of spatial feature distributions across four recognition methods (Point set: Lobectomy vdeo 77; Color: Different step annotations.) c: Color-coded ribbon illustrations for two representative cases from the robotic lobectomy dataset.

3.3 Ablation Studies

To evaluate the contribution of each component in our framework, we conduct an ablation study on the robotic lobectomy dataset, as shown in Table 3. All model variants are trained under identical settings to ensure fair comparisons.

Without the Global Temporal Mamba (GTM) module, the model performs poorly, with an accuracy of only 55.84%, highlighting the necessity of temporal modeling for step recognition. We then evaluate the effect of stacking different numbers of Local Feature Refinement (LFR) layers. Adding a single LFR layer yields slight gains, while stacking two LFR layers leads to the best overall per-

Table 3. Ablation study on the number of stacked Local Feature Refinement (LFR) layers after the baseline, Global Temporal Mamba (GTM).

GTM	LFR Layers	Accuracy(%)	Precision(%)	Recall(%)	F1-score(%)	Jaccard(%)
–	0	56.96	32.44	56.96	41.34	32.44
–	1	56.96	32.44	56.96	41.34	32.44
–	2	55.84	33.24	55.84	41.57	32.31
✓	0	90.54	89.61	90.54	89.81	84.45
✓	1	90.58	89.38	90.85	89.91	84.76
✓	2	**92.15**	**90.88**	**92.14**	**91.36**	**86.98**
✓	3	90.78	89.16	90.78	89.81	84.68

formance, with an accuracy of 92.15%, F1-score of 91.36%, and Jaccard index of 86.98%. However, adding a third LFR layer results in a marginal drop, suggesting that excessive refinement may introduce redundancy or overfitting. These results validate that both global temporal modeling and appropriately scaled local feature refinement are critical for achieving robust surgical step recognition.

4 Conclusion

In this work, we address the problem of key step recognition in laparoscopic pulmonary lobectomy, a procedure marked by high variability and anatomical complexity. We propose a framework that integrates global temporal modeling and local feature refinement to capture both long-range dependencies and fine-grained dynamics. Additionally, we construct a clinically validated dataset covering diverse lobectomy subtypes and intraoperative variations. By modeling global and local patterns from sparsely sampled frames, our approach provides a reference for workflow understanding in complex thoracic surgeries. We expect that the proposed framework and dataset could facilitate future research on full-procedure lobectomy workflow recognition, zero-shot identification of unlabeled surgical steps, and other related tasks.

Acknowledgement. The work described in this paper was supported by a grant from the Science, Technology and Innovation Commission of Shenzhen Municipality Project No. SGDX20220530111201008.

References

1. Czempiel, T., et al.: Tecno: surgical phase recognition with multi-stage temporal convolutional network. In: Medical Image Computing and Computer Assisted Intervention–MICCAI 2020: 23rd International Conference, Lima, Peru, 4–8 October 2020, Proceedings, Part III 23, Springer, pp. 343–352 (2020)
2. Gao, X., et al.: Trans-SVNet: accurate phase recognition from surgical videos via hybrid embedding aggregation transformer (2021). arXiv:2103.09712 [cs.CV]. https://arxiv.org/abs/2103.09712
3. Garrow, C.R., et al.: Machine learning for surgical phase recognition: a systematic review. Ann. Surgery **273**(4), 684–693 (2021)
4. Gu, A., Dao, T.: Mamba: linear-time sequence modeling with selective state spaces. *arXiv preprint*arXiv:2312.00752 (2023)
5. Jin, Y., et al.: Temporal memory relation network for workflow recognition from surgical video. IEEE Trans. Med. Imaging **40**(7), 1911–1923 (2021)
6. Jin, Y., et al.: Trans-SVNet: hybrid embedding aggregation Transformer for surgical workflow analysis. Int. J. Comput. Assist. Radiol. Surg. **17**(12), 2193–2202 (2022)
7. Kirtac, K., et al.: Surgical phase recognition: from public datasets to real-world data. Appl. Sci. **12**(17), 8746 (2022)

8. Lampridis, S., et al.: Robotic versus video-assisted thoracic surgery for lung cancer: short-term outcomes of a propensity matched analysis. Cancers **15**(8), 2391 (2023)
9. Li, K., et al.: Videomamba: State space model for efficient video understanding. In: European Conference on Computer Vision, pp. 237–255. Springer (2024)
10. Metchik, A., et al.: A novel approach to quantifying surgical workflow in robotic-assisted lobectomy. Int. J. Med. Robot. Comput. Assist. Surg. **20**(1), e2546 (2024)
11. Patel, Y.S., et al.: RAVAL trial: protocol of an international, multicentered, blinded, randomized controlled trial comparing robotic-assisted versus video-assisted lobectomy for early-stage lung cancer. PLoS One **17**(2), e0261767 (2022)
12. Song, Z., et al.: The learning curve on uniportal video-assisted thoracoscopic lobectomy with the help of postoperative review of videos. Front. Oncol. **13**, 1085634 (2023)
13. Twinanda, A.P., et al.: Endonet: a deep architecture for recognition tasks on laparoscopic videos. IEEE Trans. Med. Imaging **36**(1), 86–97 (2016)
14. Valderrama, N., et al.: Towards holistic surgical scene understanding. In: International Conference on Medical Image Computing and Computer-Assisted Intervention, pp. 442–452. Springer (2022)
15. Wang, Z., et al.: Autolaparo: a new dataset of integrated multi-tasks for image-guided surgical automation in laparoscopic hysterectomy. In: International Conference on Medical Image Computing and Computer-Assisted Intervention, pp. 486–496. Springer (2022)
16. Yang, S., et al.: Surgformer: surgical transformer with hierarchical temporal attention for surgical phase recognition. In: International Conference on Medical Image Computing and Computer-Assisted Intervention, pp. 606–616. Springer (2024)
17. Zhang, X., et al.: SPRMamba: surgical phase recognition for endoscopic submucosal dissection with mamba. *arXiv preprint*arXiv:2409.12108 (2024)
18. Zisimopoulos, O., et al.: DeepPhase: surgical phase recognition in CATARACTS videos. In: Frangi, A.F., Schnabel, J.A., Davatzikos, C., Alberola-López, C., Fichtinger, G. (eds.) MICCAI 2018. LNCS, vol. 11073, pp. 265–272. Springer, Cham (2018). https://doi.org/10.1007/978-3-030-00937-3_31
19. Zisserman, A., et al.: The kinetics human action video dataset. *arXiv preprint arXiv 1705* (2017)

Towards Robust Surgical Automation via Digital Twin Representations from Foundation Models

Hao Ding[✉], Lalithkumar Seenivasan, Hongchao Shu, Grayson Byrd, Han Zhang, Pu Xiao, Juan Antonio Barrag, Russell H. Taylor, Peter Kazanzides, and Mathias Unberath

Johns Hopkins University, 3400 N Charles Street, Baltimore, MD 21218, USA
{hding15,unberath}@jhu.edu

Abstract. Large language model-based (LLM) agents are emerging as a powerful enabler of robust embodied intelligence due to their capability of planning complex action sequences. Sound planning ability is necessary for robust automation in many task domains, but especially in surgical automation. These agents rely on a highly detailed natural language representation of the scene. Thus, to leverage the emergent capabilities of LLM agents for surgical task planning, developing similarly powerful and robust perception algorithms is necessary to derive a detailed scene representation of the environment from visual input. Previous research has focused primarily on enabling LLM-based task planning while adopting simple yet severely limited perception solutions to meet the needs for bench-top experiments, but lacks the critical flexibility to scale to less constrained settings. In this work, we propose an alternate perception approach – a digital twin (DT) -based machine perception approach that capitalizes on the convincing performance and out-of-the-box generalization of recent vision foundation models. Integrating our DT representation and LLM agent for planning with the dVRK platform, we develop an embodied intelligence system and evaluate its robustness in performing peg transfer and gauze retrieval tasks. Our approach shows strong task performance and generalizability to varied environmental settings. Despite a convincing performance, this work is merely a first step towards the integration of DT representations. Future studies are necessary for the realization of a comprehensive DT framework to improve the interpretability and generalizability of embodied intelligence in surgery.

Keywords: Surgical Intelligence · Medical Robotics

1 Introduction

Surgical robots, such as the da Vinci systems, offer enhanced precision, dexterity, control, and visualization, facilitating minimally invasive surgeries that result in fewer complications and faster recovery times. With research platforms like the da Vinci Research Kit (dVRK) [20] enabling the initial exploration of surgical task automation, emerging language-based automation methods [1,2,27] and policy learning methods [11,22] have further accelerated efforts in surgical task automation. Popular tasks to demonstrate surgical automation include peg transfer [11,17,18], suturing [12,19,33], knot tying [22], vascular shunt insertion [5], and needle picking [27,40] because they offer repeatable testbeds that challenge automation approaches both with respect to task planning and task execution. Large language model (LLM)-based automation [27,30,34], in particular, has recently enjoyed particular popularity because LLM agents enable long-horizon planning, potentially in an explainable and interactive way. Capitalizing on the potential benefits of LLM-based automation, however, relies on two key factors: (a) the ability to create detailed scene representations via machine perception, and (b) LLM agent setup to enable task-level planning and control. While previous approaches to LLM-based automation have started to demonstrate promising results, they mostly focus on the latter aspect, how to leverage LLM agents for advanced control planners and policy learning techniques. Robust scene representation via machine perception, however, is a critical prerequisite for LLM-based automation. In this work, we present a DT-based approach to LLM-based automation, leveraging robust vision foundation models to extract scene representation from visual input.

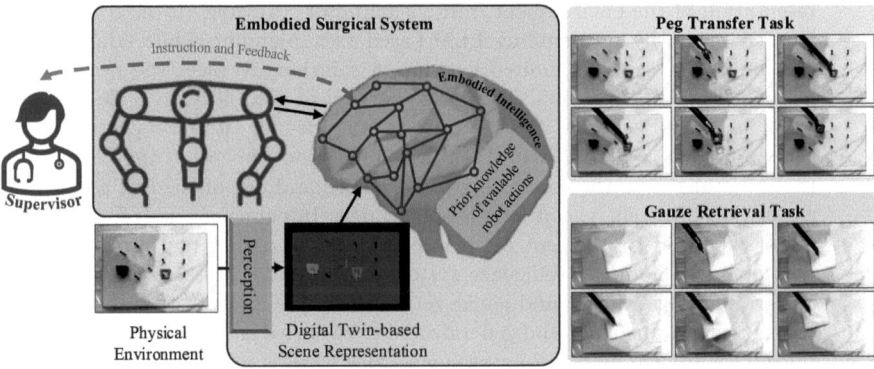

Fig. 1. Illustration of the DT-based embodied surgical system. A machine perception module is applied to extract the DT representation from the physical environment. An LLM-enabled embodied intelligence takes commands from a supervisor and makes high-level task plans based on the scene representation, prior knowledge, available actions, and previous actions and feedback. A robotic system receives commands and executes them in the physical world.

Digital twins (DT), computational replicas of the real world (physical twin) created and updated through sensor data analysis, such as machine vision, offer an intermediary layer between the low-level processes (e.g., vision tasks) and the high-level scene analysis and automation tasks. This DT-based paradigm for automation offers a unifying framework for low- and high-level analysis and automation in a more generalizable and interpretable manner [8]. To obtain the DT representation, previous works [16,21,24,36] predominantly relied on external tracking devices and markers to ensure the robustness and accuracy of the system. While advancements in deep learning algorithms for computer vision, such as instance segmentation [3,4,7,13] and pose estimation [14,15,25,26,29,37], offer an alternate vision-based, marker-less approach to extract the DT representation, these methods lack generalizability and fail when the observed scenario differs from the training data [6,9,10]. The recent emergence of vision foundation models [23,31,32,39,41,42] offers more generalizable tools for creating DT representations and developing robust machine perception [28,35]. These advancements can complement powerful LLM-based planners and robot control systems, creating a framework that affords the necessary robustness to accelerate the advancement of surgical task automation.

In this paper, we demonstrate the aforementioned concept by instantiating an embodied surgical system enabled by a basic DT representation. We propose a machine perception module to extract the DT representation robustly. As shown in Fig. 1, the perception module takes the vision input and extracts the DT representation. The representation is provided to the LLM-enabled embodied intelligence for task planning and further commands the robotic control unit for task execution. We take peg transfer and gauze retrieval as our experimental tasks. We find that our embodied surgical system presents promising automation performance in terms of success rate, exhibiting strong robustness to variations in the experimental environments where rule-based and specifically trained neural network baselines tend to fail. In summary, our key contributions are:

– Proposing a foundation model-based machine perception for extracting DT representation from the physical world.
– Proposing an embodied intelligence surgical system, enabled by the DT representation, that presents robust automation performance.

2 Method

2.1 Embodied Surgical System Overview

Our embodied surgical system incorporates three main components: DT-based machine perception, robotic control system, and embodied intelligence (language-based agent) (Fig. 1). The DT-based machine perception utilizes RGB-D data extracted from the environment to track objects of interest and generate a basic DT representation of the workspace. The robotic control system applies the da Vinci Research Kit (dVRK) [20], which facilitates the control of the surgical system's Patient Side Manipulator (PSM) to execute the planned action. Taking on the role of embodied intelligence, the language-based agent processes human-level natural language commands and generates corresponding action plans for the robot. These plans are based on the language input, the DT representation, available robot control actions, and real-time feedback.

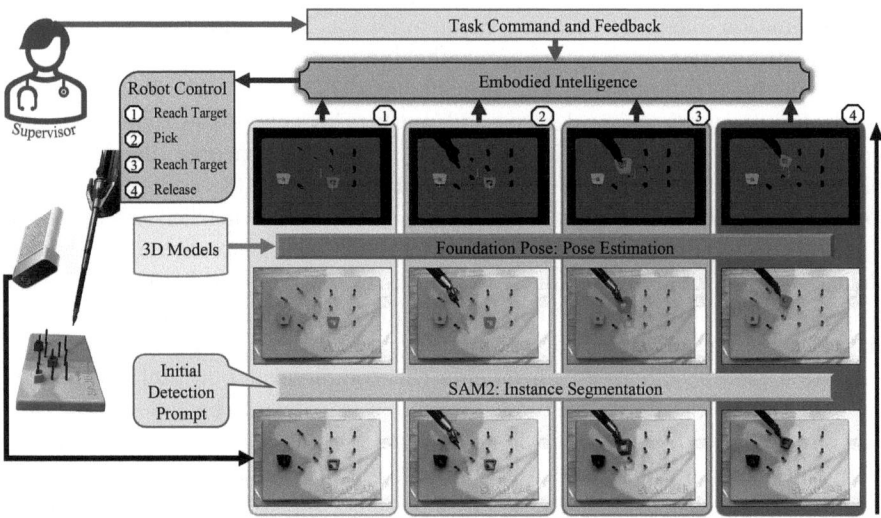

Fig. 2. Illustration of the workflow of the proposed embodied surgical system with DT-based machine perception. The captured image is first processed via SAM2 [32] with initial point prompts for the objects of interest. The objects' identification, segmentation, raw image, and corresponding 3D models are processed via the FoundationPose model to predict 6DoF poses. The extracted information forms a DT representation and is further captured by embodied intelligence for task planning.

2.2 Digital Twin-Based Machine Perception

Digital Twin Representation. The DT representation is the quantified information that can be used to construct a DT-based physical environment (physical twin). This representation can encompass identification, geometric, spatial, and physical information like label, shape, pose, and friction. In this work, we apply a basic DT representation using identification, segmentation, 3D models, and 6 DoF poses of the object of interest, which are necessary for automating basic tasks like peg transfer and gauze retrieval.

Perception Workflow. The DT-based machine perception utilizes the sensory data from an RGB-D sensor to extract the DT representation through a sequential workflow, as shown in Fig. 2. During initialization, the Segment Anything Model 2 (SAM2) [32] is prompted with points that initialize the identification and segmentation of the objects of interest. In the subsequent tracking and update phase, the objects of interest are continuously detected and tracked to update the DT representation in real-time. In this phase, the input RGB-D sensory data is first propagated through the SAM2 model to segment the objects of interest. These segments, along with the 3D model priors of the object and raw image, are then processed by the FoundationPose [39] model to extract the corresponding 6 DoF poses to form the DT representation.

2.3 Robotic Control System

The robotic control system is comprised of the da Vinci Classic surgical system (hardware) and the dVRK [20] platform (software). The surgical system includes Patient Side Manipulators (PSMs), which are controlled using the dVRK's integrated control system to execute low-level actions (e.g., measured_cp for forward kinematics, move_cp for moving in Cartesian space). Before task execution, we first perform a hand-eye calibration using the provided pipelines from dVRK to align the robot's base with the camera coordinates. We use forward kinematics to get the position and orientation in the camera coordinates.

2.4 Embodied Intelligence

A language-based agent using GPT4-o is employed to realize embodied intelligence. The system prompt defines the agent's role and provides a set of actions from which the agent can carefully select and sequence to complete a task. Based on the human-level natural language commands, the agent performs step-by-step online planning. At each step, it predicts the next action based on the task command, previous actions, and the supervisor's feedback. Here, the supervisor integrates human feedback into each step to enable closed-loop planning through shared embodiment. The set of actions made available to the agent includes perception actions and robot actions, as listed below:

1. Get observations: allows the agent to access the extracted DT representation of the environment, such as the identification, segmentation, and pose of objects. Each object is assigned an object ID to aid future planning.
2. Reach target object: enables the agent to control the surgical system's PSM to reach the pick/place position of an object with a specific object ID.
3. Pick target object: allows the agent to close the end-effector (Large Needle Driver) attached to the PSM, to grab/pick the target object.
4. Release the object: allows the agent to open the end-effector, releasing a picked object at the current position.
5. Adjust position: allows the agent to incrementally adjust the robot's position by a fixed offset relative to the camera coordinates based on the specified directions: up, down, left, right, forward, and back.
6. Inquiry: allows the agent to interact with the supervisor to get further instructions or clarifications.

After completing a reach/pick action (2, 3), the agent requests feedback from the supervisor to confirm the successful execution of the action. During the reach actions (2), the PSM follows a trajectory based on linearly interpolated waypoints decided from the current and final positions.

3 Experiment

We employ the Azure Kinect RGB-D camera as the vision sensor for machine perception and the dVRK system as the robotic control system [20] in our embodied surgical system. We benchmark our system against two baseline methods

(Sect. 3.1) on the peg transfer task, a common laparoscopic training task used for skill training and assessment in surgical training programs. Additionally, the task generalizability of each system is assessed using a gauze retrieval task.

Fig. 3. Comparison between our perception module and baseline models.

3.1 Baseline Methods

We applied two variants as our baseline models. The comparison between our model and baseline models are shown in Fig. 3 **Depth thresholding (DTh) + Iterative Closest Point (ICP):** We adopt the depth thresholding + ICP method from Hwang et al. [17,18]. We threshold both the upper bound and the lower bound of the depth to get the target object. The threshold is calculated as $[min(d_{positive}) - \epsilon_{lb}, min(d_{negative}) - \epsilon_{ub}]$, where $d_{positive}$ and $d_{negative}$ are the depths for positive and negative prompt points, and (ϵ_{lb}) and (ϵ_{ub}) are the lower and upper bound depth noise tolerance for effective target object-background separation. We initialize translation and rotation to the pose of the back-projected point from the center pixel of the object and identity matrix, respectively. We then apply ICP to refine the final pose of the objects using the projected points and the 3D models.

YOLO + ICP: We incorporates YOLOv8 [38] and ICP. We custom-trained YOLOv8 for instance segmentation on data collected from the ideal experiment setup, with annotation generated by SAM2 and filtered by human annotators. To simulate the initial point prompts provided to the SAM2 model in the other baseline and our method, visible points are added to the images, with distinct colors indicating different objects, for both training and inference. It applies the same pose estimation method using ICP.

3.2 Peg Transfer in Varied Environments

We evaluate the robustness of our embodied surgical system, driven by a digital scene representation derived from foundation models, on a peg transfer task. The task involves a pegboard with 12 pegs and some blocks initially placed on the pegs. The robot must pick a specific block and place it on a target peg. One pick-and-place action sequence is considered as one trial. We benchmark our system on both open-loop and closed-loop planning to disentangle the advantages of robust language agents and highlight the effectiveness of our DT-based machine perception. In the open-loop planning framework, the agent plans the actions, and the robotic control system executes them once without any supervisor feedback to verify successful action completion. In the closed-loop planning framework, the agent accepts language feedback from the supervisor. This feedback includes fine-grained position adjustment of the robot end-effector in six directions (up, down, left, right, forward, and back) in the image space, target re-detection, and re-execution of the action. Each position adjustment feedback will adjust the end-effector tool tip position by 3 mm in the specified direction in camera coordinates. A maximum of 5 position adjustments or redos is allowed for each trial before considering it a failure trial. Both open and closed-loop planning frameworks are evaluated based on the success rate and the failure modes: inaccurate pose (Po), object not detected (De), and planning error (Pl). The closed-loop framework is also evaluated on the number of planning steps.

Table 1. Experiment Results

Experimental Setup	Method	Closed-loop planning			Open-loop planning	
		Success Rate	Average Planning Steps	Failure Mode Po, De, Pl	Success Rate	Failure Mode Po, De, Pl
Ideal Environment	DTh + ICP	97% (97/100)	5.59	1, 2, 0	73% (73/100)	25, 2, 0
	YOLO + ICP	97% (97/100)	5.64	3, 0, 0	75% (75/100)	25, 0, 0
	Ours	**100% (100/100)**	**5.04**	**0, 0, 0**	**96% (96/100)**	**4, 0, 0**
Black/Red Block	DTh + ICP	88% (44/50)	5.80	3, 3, 0	46% (23/50)	24, 3, 0
	YOLO + ICP	72% (36/50)	5.36	8, 6, 0	54% (27/50)	17, 6, 0
	YOLO + FP	90% (45/50)	5.04	0, 5, 0	86% (43/50)	2, 5, 0
	Ours	**100% (50/50)**	**5.08**	**0, 0, 0**	**96% (48/50)**	**2, 0, 0**
Tilted Pegboard	DTh + ICP	56% (28/50)	6.79	11, 10, 1	8% (4/50)	35, 10, 1
	DTh + FP	78% (39/50)	5.23	1, 10, 0	68% (34/50)	6, 10, 0
	YOLO + ICP	84% (41/50)	6.00	7, 2, 0	36% (18/50)	30, 2, 0
	Ours	**96% (48/50)**	**5.10**	**2, 0, 0**	**86% (43/50)**	**7, 0, 0**
Gauze Retrival	Depth thresholding + ICP	84% (84/100)	-	15, 0, 1	-	-
	YOLO + ICP (Peg transfer data)	0% (0/100)	-	0, 100, 0	-	-
	YOLO + ICP (Gauze data)	100% (100/100)	-	0, 0, 0	-	-
	Ours	100% (100/100)	-	0, 0, 0	-	-

Varied Environments. Our embodied surgical system, in both its open and closed-loop planning configurations, is evaluated against the two baseline models on three varied environments to evaluate the effectiveness of the foundation model-enabled DT representation.

These environments include:

- Ideal environment: The pegboard is positioned at the center of the camera's field of view, with its normal direction perpendicular to the camera plane. We use the grey trapezoid block.
- Changing block color: The block color is changed to black and red.
- Changing pegboard orientation: The pegboard is tilted at a fixed angle ($\approx$ 15°) toward the camera plane.

Each method, in each of its planning frameworks, is tested over 100 trials in the ideal environment and 50 trials for each varied environment.

Results and Discussion. Table 1 quantitatively benchmarks our method against the two baseline methods on varied environments, in both the closed and open-loop planning frameworks. In the closed-loop framework, while all methods achieved a high success rate in the ideal environment, a drop in success rate and an increase in average planning steps are observed for the two baseline models in the other two environments. This indicates that the two baseline methods' machine perceptions are less robust, as indicated by the increase in pose estimation and detection error observed in the respective failure modes. The rise in average planning steps for the baseline models suggests that embodied intelligence is attempting to compensate for the limitations of its machine perception.

The flexibility and generalizability of our DT-based machine perception, which leverages a foundation model, becomes much more evident when we disentangle (open-loop planning framework) the robustness of embodied intelligence. In the open-loop planning framework, our method outperforms the baseline methods under the ideal environment and varied environments. The limited flexibility of the two baseline methods can be attributed to several factors. The effectiveness of the depth thresholding technique is primarily limited by two main factors: (a) the black color absorbs infrared light, which interferes with depth estimation, and (b) the tilted pegboard makes it harder to threshold the depths between the board and the block. The YOLO model struggled primarily due to the out-of-domain predictions, as the black/red block and tilted pegboard were not included in the training set. As a result, the YOLO model either fails to detect the object or predicts inaccurate segmentation.

Ablation Studies. Additionally, we perform ablation studies in varied environments to explore the effectiveness of each component and design choices. We replace ICP with FoundationPose (FP) for YOLO for the black/red blocks environment and for Depth thresholding in the tilted pegboard environment. We use the same setting for varied environments. Results in Table 1 show that, although the detection failure cannot be addressed, FoundationPose alleviates the pose estimation error caused by inaccurate segmentation with visual input.

3.3 Task Generalization: Gauze Retrieval

To further validate the generalizability of our embodied surgical system, we evaluate its performance on gauze retrieval tasks. This task requires the end-effector to pick up a 5 cm × 5 cm gauze, with each pick-up action considered as a single trial. All methods are evaluated in an open-loop planning framework, based on the success rate in 100 trials.

From the quantitative results in Table 1, we observe that our method achieves robust performance, demonstrating its zero-shot generalization ability for this task. In contrast, the performance of the baseline method employing the depth thresholding technique declines due to the inseparable depth between the gauze and the background. Similarly, the YOLOv8-based method initially failed to complete the task even once due to the out-of-domain challenges. However, when the YOLOv8 model is further trained on an additional 100 images with gauze annotations, its performance improves to levels comparable with our method.

4 Conclusion

With most research on embodied intelligence focusing mainly on advancing language-based agents for robust task planning, we propose an alternate approach focusing on advancing machine perception. We leverage foundation models to extract DT representation to serve as an intermediary layer to complement the LLM-based embodied intelligence and create a flexible, scalable, interpretable, and generalizable surgical embodied system. Our instantiation for the surgical training tasks showcases its potential for robust task automation.

Besides this, a more comprehensive DT framework allows the generation of massive synthetic data to train high-level scene analysis and automation agents. This, in turn, could potentially enhance the adaptability and generalizability of embodied intelligence in surgery. Challenges remain when apply our idea into soft-tissue manipulation as it requires high-fidelity construction of digital twin representation which requires precise calibration, extensive data integration, and consistent visual fidelity. Thus, future advancements are required in perception, modeling, and simulation in soft-tissue digital twin and more efforts are expected to explore the DT-driven approaches' potential in this regard to advance surgical automation, moving it closer to practical, real-world clinical applications.

Acknowledgments. his research is supported by a collaborative research agreement with the MultiScale Medical Robotics Center at The Chinese University of Hong Kong.

Disclosure of Interests. The authors have no competing interests to declare that are relevant to the content of this article.

References

1. Brohan, A., et al.: Rt-1: robotics transformer for real-world control at scale. arXiv preprint arXiv:2212.06817 (2022)
2. Brohan, A., et al.: Rt-2: vision-language-action models transfer web knowledge to robotic control. arXiv preprint arXiv:2307.15818 (2023)
3. Chen, K., et al.: Hybrid task cascade for instance segmentation. In: Proceedings of CVPR, pp. 4974–4983 (2019)
4. Cheng, B., et al.: Masked-attention mask transformer for universal image segmentation. In: Proceedings of CVPR, pp. 1290–1299 (2022)
5. Dharmarajan, K., et al.: Automating vascular shunt insertion with the dvrk surgical robot. In: Proceedings of ICRA, pp. 6781–6788. IEEE (2023)
6. Ding, H., et al.: SegSTRONG-C: segmenting surgical tools robustly on non-adversarial generated corruptions – an endovis'24 challenge (2024)
7. Ding, H., Qiao, S., Yuille, A., Shen, W.: Deeply shape-guided cascade for instance segmentation. In: Proceedings of CVPR, pp. 8278–8288 (2021)

8. Ding, H., Seenivasan, L., Killeen, B.D., Cho, S.M., Unberath, M.: Digital twins as a unifying framework for surgical data science: the enabling role of geometric scene understanding. ais **4**(3), 109–138 (2024)
9. Ding, H., Wu, J.Y., Li, Z., Unberath, M.: Rethinking causality-driven robot tool segmentation with temporal constraints. Int. J. CARS 1009 – 1016 (2022)
10. Ding, H., Zhang, J., Kazanzides, P., Wu, J.Y., Unberath, M.: CaRTS: causality-driven robot tool segmentation from vision and kinematics data. In: Proceedings of MICCAI, pp. 387–398. Springer (2022)
11. Fu, J., Long, Y., Chen, K., Wei, W., Dou, Q.: Multi-objective cross-task learning via goal-conditioned GPT-based decision transformers for surgical robot task automation. arXiv preprint arXiv:2405.18757 (2024)
12. Hari, K., et al.: STITCH: augmented dexterity for suture throws including thread coordination and handoffs. arXiv preprint arXiv:2404.05151 (2024)
13. He, K., Gkioxari, G., Dollár, P., Girshick, R.: Mask r-cnn. In: Proceedings of ICCV, pp. 2961–2969 (2017)
14. He, Z., Feng, W., Zhao, X., Lv, Y.: 6d pose estimation of objects: recent technologies and challenges. Appl. Sci. **11**(1), 228 (2020)
15. Hein, J., et al.: Towards markerless surgical tool and hand pose estimation. Int. J. Comput. Assist. Radiol. Surg. **16**(5), 799–808 (2021). https://doi.org/10.1007/s11548-021-02369-2
16. Hein, J., et al.: Creating a digital twin of spinal surgery: a proof of concept. In: Proceedings of CVPR, pp. 2355–2364 (2024)
17. Hwang, M., et al.: Automating surgical peg transfer: calibration with deep learning can exceed speed, accuracy, and consistency of humans. IEEE Trans. Autom. Sci. Eng. **20**(2), 909–922 (2022)
18. Hwang, M., et al.: Efficiently calibrating cable-driven surgical robots with RGBD fiducial sensing and recurrent neural networks. IEEE RAL **5**(4), 5937–5944 (2020)
19. Kam, M., et al.: Autonomous system for vaginal cuff closure via model-based planning and markerless tracking techniques. IEEE RAL **8**(7), 3916–3923 (2023)
20. Kazanzides, P., et al.: An open-source research kit for the da Vinci® surgical system. In: Proceedings of ICRA, pp. 6434–6439. IEEE (2014)
21. Killeen, B.D., et al.: Stand in surgeon's shoes: virtual reality cross-training to enhance teamwork in surgery. Int. J. CARS 1–10 (2024)
22. Kim, J.W., et al.: Surgical robot transformer (srt): imitation learning for surgical tasks. arXiv preprint arXiv:2407.12998 (2024)
23. Kirillov, A., et al.: Segment anything. In: Proceedings of ICCV, pp. 4015–4026 (2023)
24. Kleinbeck, C., Zhang, H., Killeen, B.D., Roth, D., Unberath, M.: Neural digital twins: reconstructing complex medical environments for spatial planning in virtual reality. Int. J. CARS **19**(7), 1301–1312 (2024)
25. Li, Z., et al.: Tatoo: vision-based joint tracking of anatomy and tool for skull-base surgery. Int. J. CARS **18**(7), 1303–1310 (2023)
26. Marullo, G., et al.: 6d object position estimation from 2d images: a literature review. Multimedia Tools Appl. **82**(16), 24605–24643 (2023)
27. Moghani, M., et al.: SuFIA: language-guided augmented dexterity for robotic surgical assistants. arXiv preprint arXiv:2405.05226 (2024)
28. Oguine, K.J., Mukul, R.D.S., Drenkow, N., Unberath, M.: From generalization to precision: exploring SAM for tool segmentation in surgical environments. In: Medical Imaging 2024: Image Processing, vol. 12926, pp. 7–12. SPIE (2024)
29. Peng, S., Liu, Y., Huang, Q., Zhou, X., Bao, H.: Pvnet: pixel-wise voting network for 6dof pose estimation. In: Proceedings of CVPR, pp. 4561–4570 (2019)

30. Qin, Y., et al.: ToolLLM: facilitating large language models to master 16000+ real-world APIs (2023)
31. Raiciu, C., Rosenblum, D.S.: Enabling confidentiality in content-based publish/subscribe infrastructures. In: Securecomm and Workshops, pp. 1–11 (2006)
32. Ravi, N., et al.: Sam 2: segment anything in images and videos. arXiv preprint arXiv:2408.00714 (2024)
33. Saeidi, H., et al.: Autonomous robotic laparoscopic surgery for intestinal anastomosis. Sci. Robot. **7**(62), eabj2908 (2022)
34. Schick, T., et al.: Toolformer: language models can teach themselves to use tools (2023)
35. Shen, Y., Ding, H., Shao, X., Unberath, M.: Performance and non-adversarial robustness of the segment anything model 2 in surgical video segmentation. arXiv preprint arXiv:2408.04098 (2024)
36. Shu, H., et al.: Twin-S: a digital twin for skull base surgery. Int. J. CARS **18**(6), 1077–1084 (2023)
37. Teufel, T., et al.: OneSLAM to map them all: a generalized approach to SLAM for monocular endoscopic imaging based on tracking any point. Int. J. CARS 1–8 (2024)
38. Varghese, R., Sambath, M.: YOLOv8: a novel object detection algorithm with enhanced performance and robustness. In: Proceedings of ADICS, pp. 1–6. IEEE (2024)
39. Wen, B., Yang, W., Kautz, J., Birchfield, S.: Foundationpose: unified 6d pose estimation and tracking of novel objects. In: Proceedings of CVPR, pp. 17868–17879 (2024)
40. Wilcox, A., et al.: Learning to localize, grasp, and hand over unmodified surgical needles. In: Proceedings of ICRA, pp. 9637–9643. IEEE (2022)
41. Xiao, Y., et al.: Spatialtracker: tracking any 2d pixels in 3d space. In: Proceedings of CVPR, pp. 20406–20417 (2024)
42. Yang, L., et al.: Depth anything: Unleashing the power of large-scale unlabeled data. In: Proceedings of CVPR, pp. 10371–10381 (2024)

Semantic Scene Editing for Cholecystectomy Surgery

Çağhan Köksal[1(✉)], Yousef Yeganeh[1,2], Nassir Navab[1,2], and Azade Farshad[1,2]

[1] Technical University of Munich, Munich, Germany
caghan.koksal@tum.de
[2] Munich Center for Machine Learning, Munich, Germany

Abstract. Semantic scene editing in the surgical domain presents unique challenges due to the need to preserve anatomical fidelity while altering specific scene elements. In this paper, we propose a novel image editing framework for cholecystectomy surgery using diffusion models. Our approach enables targeted modifications of surgical scenes—such as tool removal, relocation, rotation, and replacement—while maintaining a coherent representation of the operative field. By leveraging the conditional control capabilities of the diffusion model, our model semantically understands the surgical context and performs realistic inpainting to generate high-fidelity edited images. Our comprehensive quantitative and qualitative evaluations on the Cholec dataset demonstrate the proposed model's superiority and effectiveness in preserving structural details and ensuring visual consistency in the scene editing task.

Keywords: Scene Synthesis · Semantic Editing · Surgical Videos

1 Introduction

Surgical scene understanding and manipulation have emerged as fundamental challenges in computer-assisted interventions, with applications spanning surgical training, intraoperative guidance, and robotic surgery [1]. While recent advances in surgical video analysis have enabled action recognition [2–4] and workflow understanding [5–7], the ability to semantically edit surgical scenes remains largely unexplored. Such capability is crucial for data augmentation in low-data regimes [8], creating diverse training scenarios, and developing surgical simulators that can adapt to specific procedural variations. Existing approaches to surgical data generation have primarily focused on synthesizing entirely new content: unconditional video generation [9], text-to-image synthesis [10], graph-to-image synthesis [11], and action-conditioned video generation [12,13]. While these methods demonstrate the potential of generative models in surgery, they lack the fine-grained control necessary for targeted scene modifications—a capability essential for creating specific surgical scenarios, correcting annotations, or augmenting rare events. Recent breakthroughs in diffusion models have revolutionized image editing in natural scenes. Stable Diffusion [14] and its inpainting variants enable high-fidelity local modifications, while frameworks like ControlNet [15] provide conditional control over generation. While these innovations, exemplified by frameworks such as EditAnything [16], have broadened the horizon for general-purpose image editing, extending these techniques to the surgical domain involves additional complexities. In surgery, the integration of modified content with clinically relevant anatomical details is paramount, thereby necessitating specialized adaptations of these techniques. Directly applying these methods to surgical scenes yields anatomically implausible results due to their lack of understanding of surgical semantics, tool-tissue interactions, and procedural constraints. The surgical domain demands preservation of critical anatomical structures, realistic tissue deformation, and clinically valid tool positioning. The evolution from GANs [17,18] to diffusion models [19,20] in medical image generation has demonstrated superior visual quality and training stability. Recent surgical works have explored this potential using guided generation for segmentation tasks [21], or concept erasing in surgical contexts [22]. However, these approaches either focus on global generation or simple removal tasks, lacking the capability for complex semantic manipulations such as tool replacement or repositioning while maintaining scene coherence. Our work builds upon these advances by integrating a diffusion-based inpainting module with conditional control mechanisms tailored for surgical scene modifications. By leveraging the context-aware abilities of diffusion models, our framework achieves realistic alterations in regions such as surgical tool placements while ensuring the integrity of surrounding anatomical structures. This approach not only benefits from the robust synthesis capabilities of modern generative models but also addresses the unique challenges posed by the surgical environment.

To summarize, our contributions are: (1) we propose an image editing framework tailored for surgical scene manipulation, (2) we introduce a semantic control mechanism that enables fine-grained editing (translation, rotation, tool replacement, and tool removal) based on triplet prompts and binary tool masks, (3) we provide extensive qualitative and quantitative evaluations demonstrating the effectiveness of our approach across various surgical scenarios, and (4) we release project materials at https://caghankoksal.github.io/cholec-editing/.

2 Method

In this work, we propose an image editing framework conditioned on surgical action triplets. Our framework is designed to enable semantic editing of surgical scenes by selectively modifying regions while preserving global coherence. The editing process is driven by two diffusion models, one for tool removal and another with inpainting capabilities. Given an input RGB image $I \in \mathbb{R}^{3 \times H \times W}$ with height H and width W, a user-specified binary mask $M \in \mathbb{R}^{1 \times H \times W}$ that highlights the surgical tool or the region to be edited, and the corresponding action triplet τ, a conditional diffusion model is tasked with reconstructing the masked region such that it seamlessly integrates with the rest of the image. $I_M = I \odot M$ is the masked input image, and the reconstructed image is denoted by I'.

Tool Removal. We employ parameter-free Attentive Eraser model [23] to first perform targeted tool removal and then do inpainting within complex scenes. The removal model takes as input the image I, the binary mask M indicating the region to erase, and learned attention maps A and produces the final image $\hat{I}$ with the specified region seamlessly filled with background based on a context.

Text Embeddings. Each image I_i is associated with a set of actriplets $\tau_i = \{\tau_{i_1}, \tau_{i_2}, \ldots, \tau_{i_N}\}$, where N_i is the number of triplets for image i. Each triplet $\tau_{i,j} = (s, p, o)$ consists of three textual components, where the subject (s), predicate (p), and object (o) refer to the instrument, verb, and target, respectively. Text embeddings γ_i are obtained by embedding the triplets using the text encoder ω, where $\gamma_i = \omega(\tau_i)$ and τ_i is obtained by concatenating $\tau_{i_1}, \ldots, \tau_{i_N}$. Finally, the diffusion model is conditioned on the embedded triplets. We utilize two distinct text encoders to embed the triplets: (1) CLIP [24] and (2) SurgVLP [25], which is based on the BERT [26] architecture.

Diffusion-Based Inpainting Module. We build upon state-of-the-art (SOTA) diffusion models by incorporating conditional inpainting. The overall process can be summarized in the following steps:

- Pre-processing: Input image I is paired with a binary mask M that denotes the region targeted for editing. This mask can be based on ground truth (GT), generated manually by a user, or through an automated segmentation module.
- Conditional Diffusion Process: The diffusion model processes the masked image, utilizing its conditional control mechanism to guide the generative process. The model infers the missing content based on the surrounding anatomical context while respecting the boundaries defined by M.

Formally, the training objective can be expressed as:

$$L_{total} = L_{\text{diffusion}} + \lambda_{rec} \cdot L_{rec} \tag{1}$$

where $L_{\text{diffusion}}$ corresponds to the score matching loss employed in the diffusion process; L_{rec} is an $L1$ loss that minimizes reconstruction errors; and λ_{rec} is a hyperparameter controlling the contributions of the reconstruction term.

Conditional Control Mechanism. A critical component of our framework is its ability to semantically understand the surgical context. Using the conditional control capabilities of the diffusion model, the editing process adapts based on the spatial semantics provided by the mask M. This ensures that any alterations, such as tool replacement or repositioning, are made with respect to the underlying scene structure, leading to a realistic appearance.

Semantic Image Manipulation. Our method is capable of generating both realistic and unrealistic synthetic surgical scenes by utilizing a conditional diffusion inpainting model. The input to the diffusion model is constructed by concatenating the latent representations of the original and masked images along with the mask. To control the influence of the mask on the generated output, we introduce a mask strength coefficient λ_M. The final input to the diffusion model, denoted as Z, is obtained by concatenating the latent representation of the original image $Z_o = \text{VAE}(I)$, the masked image latent representation $Z_M = \text{VAE}(I_M)$, and the mask $\lambda_M M$, where $Z_o, Z_m \in \mathbb{R}^{B \times 4 \times H \times W}$, and $Z \in \mathbb{R}^{B \times 9 \times H \times W}$. Here, B denotes the batch size. Feeding the GT segmentation masks directly as input to the model caused overfitting to the shape of the masks and ignoring the text triplet conditions, due to the highly detailed structure of the masks. The mask strength coefficient $\lambda_M \in [0, 1]$ modulates the mask intensity, allowing for a smooth transition between weak and strong mask effects. Setting $\lambda_M = 0$ removes the mask's influence, while $\lambda_M = 1$ retains the full effect of the mask.

Mask Weighting. To mitigate the overfitting issue, we leveraged a progressive degradation strategy controlled by a parameter $s \in [0, S_{\max}]$. At $s = 0$, the original mask is used without modification. At $s = S_{\max}$, a coarse approximation is used, represented by the bounding box of the object. For intermediate values, the binary mask is smoothed using a Gaussian filter with parameters scaled linearly by s, followed by thresholding to maintain binarization [27]. This enables a smooth transition from precise to coarse annotations and allows model robustness assessment under varying annotation quality.

3 Experiments and Results

3.1 Experimental Setup

Datasets. In our experiments, we used intersecting videos from the CholectT50 [3] and CholecInstanceSeg [28] datasets, employing the same data split. Since the test splits have not been officially released, we report results on the validation set. To pre-process the data, we generate random square crops of size 512 pixels, normalize the pixel distribution to the range $[-1, 1]$, and merge all regions containing tool annotations into a single mask to train the inpainting model [29].

Implementation Details. We used the Stable Diffusion (SD) 2 inpainting model as a base diffusion model. All diffusion inpainting models were fine-tuned for 10K steps with a learning rate of 5e−6. We set the guidance scale to 15 during inference. For the text encoder, we used CLIP during training, and we leveraged both CLIP and SurgVLP during evaluation of the semantic editing capabilities.

Evaluation Protocol. To validate the visual quality of our generated images, we use three standard metrics: Fréchet Inception Distance (FID), Kernel Inception Distance (KID), and Learned Perceptual Image Patch Similarity (LPIPS). We compute these metrics for the whole image, as well as on the edited region only, denoted by *RoI* or region of interest. To evaluate the semantic editing capabilities of our model, we employed an off-the-shelf triplet detector, Rendezvous [3], and computed several evaluation metrics. These include Instrument (I), Verb

Table 1. Triplet detection performance (mAP) in inpainting and editing tasks. **I**: Instrument, **V**: Verb, **T**: Target.

Model	I ↑	V ↑	T ↑	IV ↑	IT ↑	IVT ↑
Inpainting Task						
Stable Diffusion 2 [14]	70.13	69.02	69.59	50.70	54.50	57.46
Ours	**93.29**	**89.50**	**81.79**	**71.94**	**74.79**	**77.40**
Editing Task						
Stable Diffusion 2 [14]	16.56	40.81	**49.75**	10.35	10.33	10.06
Ours	**18.71**	**43.95**	47.24	**10.67**	**13.43**	**14.48**

Table 2. Semantic and visual evaluation results in the editing task using CLIP [24] and SurgVLP [25] on the region of interest (RoI).

Model	CLIP			SurgVLP		
	Acc@1 ↑	Acc@2 ↑	Acc@3 ↑	Acc@1 ↑	Acc@2 ↑	Acc@3 ↑
Stable Diffusion 2 [14]	19.61	41.50	61.82	6.37	16.99	37.67
Ours	**42.63**	**59.49**	**72.09**	**32.93**	**52.47**	**71.31**

(V), Target (T), and Instrument-Verb-Target (IVT). The metrics measure the average precision (AP) for correctly predicting the surgical instrument (I), the action performed (V), and the target object or region (T). The IVT metric combines all three components to compute the AP for correctly predicting the entire triplet. These metrics are computed using the ivtmetrics library, providing both class-wise AP and the overall mean AP (mAP). The input of the triplet detector was resized to 256 × 448 for consistency in evaluation. We designed two evaluation scenarios. In the first, for each edited image, we extracted a cropped region corresponding to the input binary segmentation mask and computed its embedding. This was then compared to the embeddings of all possible tool classes. We report whether the class specified in the editing prompt ranked in the top-1, top-2, or top-3 most similar embeddings, denoted as @1, @2, and @3, respectively. In the second scenario, we repeat the experiment using the entire (uncropped) image.

3.2 Results

Quantitative Results. To quantitatively assess the semantic consistency of the inpainted scenes, we evaluate reconstructed images using Rendezvous [3] as a triplet detector, which predicts triplets from surgical images. We report Mean Average Precision (mAP) for individual elements (I, V, T) as well as for their combinations (IV, IT, IVT) in Table 1. We show that our method significantly outperforms the DreamBooth-based SD2 baseline across all triplet configurations. Notably, for the full triplet (IVT), our model achieves an mAP

Table 3. Quantitative visual evaluation in the reconstruction and editing tasks.

Model	FID ↓	KID ↓	LPIPS ↓	FID ↓	KID ↓	LPIPS ↓
	Whole Image			RoI		
Inpainting Task						
Stable Diffusion 2 [14]	41.07	$0.102_{\pm 0.011}$	0.0205	95.11	$0.104_{\pm 0.010}$	**0.0012**
Ours	**31.33**	$0.102_{\pm 0.012}$	**0.0140**	**68.01**	$0.102_{\pm 0.012}$	0.0167
Editing Task						
Stable Diffusion 2 [14]	**42.54**	$0.126_{\pm 0.014}$	0.0025	132.58	$0.123_{\pm 0.012}$	**0.0078**
Ours	44.31	$\mathbf{0.035_{\pm 0.007}}$	**0.0024**	**97.32**	$\mathbf{0.035_{\pm 0.008}}$	0.0476

of 77.40 compared to 57.46 for SD2, indicating better preservation of complex scene semantics. Performance gains are also observed for individual and partial combinations, suggesting that our model not only produces visually plausible outputs but also encodes accurate semantic relationships. These results confirm the effectiveness of our approach in generating anatomically and semantically faithful surgical scenes that align with clinical triplet-based annotations. In Table 2, we quantitatively evaluate the semantic and visual consistency of the edited regions. For this, we first crop the edited area based on the input inpainting region mask. We then extract visual embeddings from the cropped regions using the CLIP [24] and SurgVLP [25] models. Similarly, tool classes are converted into their corresponding embeddings. We compute similarity scores between the generated region embeddings and the tool class embeddings to assess the semantic alignment and visual fidelity of the edits. The metrics @1, @2, and @3 indicate whether the correct tool class appears among the top-1, top-2, or top-3 most similar class embeddings, respectively. We show that, for both CLIP and SurgVLP embeddings, our model produces edits that are more semantically coherent with surgical tool classes, as reflected in higher top-k retrieval scores. Table 3 presents the quantitative results for reconstructing the original triplet from the original input mask, evaluating the fidelity and semantic correctness of edited images. Our method consistently outperforms the SD2 baseline. Specifically, we achieve significant improvements in both global and region-of-interest (RoI) FID scores, with reductions from 41.07 to 31.33 and from 95.11 to 68.01, respectively, indicating more realistic and semantically faithful reconstructions. While the overall KID score remains comparable across both models, our approach notably reduces the KID_{RoI} from 0.104 to 0.102, highlighting better local coherence. Furthermore, our model achieves a substantially lower LPIPS score (0.0140 vs. 0.0205), reflecting higher perceptual similarity to the ground truth. These results underscore the ability of our method to generate visually plausible and semantically consistent edits, even in challenging surgical contexts.

Qualitative Results. Figure 1 presents a visual comparison of inpainting results between our model and the SD2 baseline [29], evaluated across a range of complex surgical scenarios. The results in columns one and three demonstrate

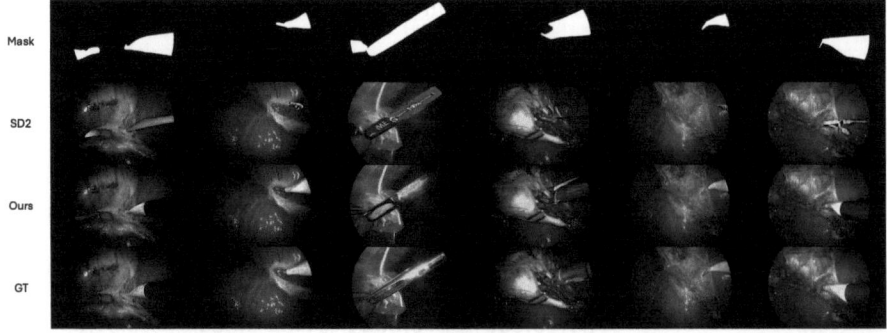

Fig. 1. Qualitative results on the inpainting task.

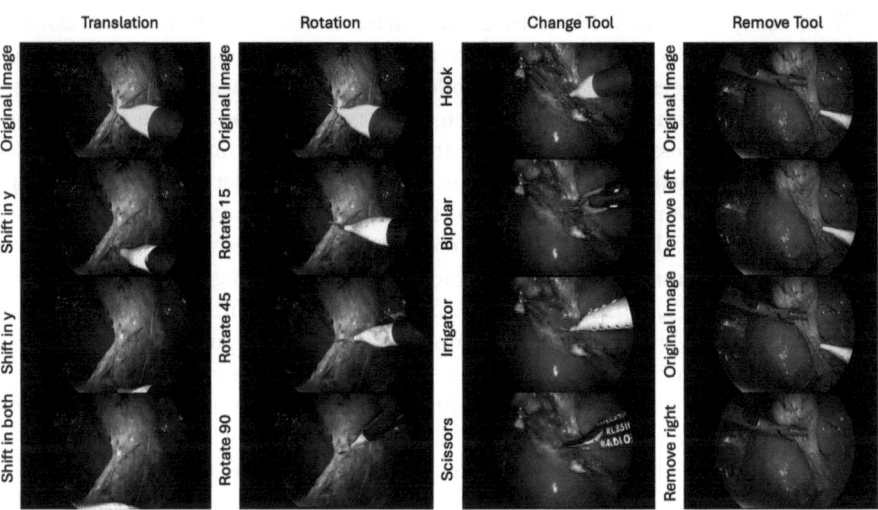

Fig. 2. Qualitative results from our image editing experiments, illustrating the model's ability to synthesize and modify surgical scenes in a controlled manner.

that our model can synthesize scenes containing multiple surgical tools. The second and fifth columns show that the model successfully generates scenes from small input masks. The third column further illustrates the model's robustness to large mask sizes. The fourth column highlights the model's ability to generate visually complex tools, while the last two columns demonstrate that the model can accurately reconstruct the correct tools even in visually challenging scenes, such as those with smoke. The SD2 baseline often introduces visual artifacts or produces semantically inconsistent outputs, particularly in cases involving multiple instruments, occlusions, or ambiguous contexts. In contrast, our model consistently generates anatomically plausible reconstructions, even under challenging conditions. It handles both small and large missing regions with robustness and accurately reconstructs fine-grained tool structures, preserving spatial and semantic coherence. These results highlight the model's ability to adapt to diverse inpainting contexts while maintaining clinical realism.

To further evaluate the model's controllability and its potential for surgical scene understanding, we conducted a set of image editing experiments, illustrated in Fig. 2. These tasks simulate clinically relevant manipulations, including tool translation, rotation, replacement, and removal (columns 1, 2, 3, 4 in Fig. 2, respectively). Our model is capable of repositioning and reorienting tools while preserving their visual identity and ensuring consistency with surrounding anatomical structures. In Fig. 2, the first row shows the original image, and the subsequent rows apply shifts of 100 and 150 pixels in the negative y-direction, and 50 pixels in the negative x-direction. The model also supports object-level edits, such as replacing one tool with another or removing a tool entirely, without compromising the realism of the scene. In the tool rotation results, tools are rotated by $15°$, $45°$, and $90°$ while maintaining structural and contextual consistency. In the third column, the hook is replaced with different instruments (bipolar, forceps, irrigator, scissors), showcasing the model's ability to integrate novel objects into the scene. Finally, in the fourth column, the tool removal task can be seen while preserving the surrounding anatomical context. These examples demonstrate the model's capacity for fine-grained and semantically coherent surgical scene editing, as well as the model's generalization capability in diverse manipulation settings, while retaining structural coherence across edits. These capabilities make the model well-suited for practical applications in data augmentation, surgical training environments, and simulation. The ability to perform targeted, semantically meaningful edits is particularly valuable for generating diverse surgical scenes and supporting downstream tasks such as tool tracking, scene understanding, and human-in-the-loop annotation.

4 Conclusion

This work introduces a high-quality semantic surgical-scene editing framework for cholecystectomy procedures. Our approach enables precise translation, rotation, replacement, and removal of instruments in real laparoscopic footage, supporting scalable synthesis of novel surgical scenes. By combining and manipulating discrete tool-related attributes, our model preserves anatomical consistency and visual realism. Furthermore, automatic generation of labeled tool masks eliminates the need for additional data collection and substantially reduces manual annotation effort. Quantitative evaluations demonstrate that the model achieves high fidelity in tool localization and appearance while preserving anatomical integrity and introducing minimal editing artifacts. Qualitative results confirm its versatility across diverse surgical contexts, enabling on-demand creation of tailored training data and facilitating downstream tasks such as instrument detection, action recognition, and workflow analysis. Overall, the proposed framework represents a significant advance in flexible, high-quality surgical scene synthesis with the potential to accelerate research and development in surgical data science and to enhance simulation-based training for surgeons.

References

1. Maier-Hein, L., et al.: Surgical data science-from concepts toward clinical translation. Med. Image Anal. **76**, 102306 (2022)
2. Nwoye, C.I., et al.: Recognition of instrument-tissue interactions in endoscopic videos via action triplets. In: Martel, A.L., et al. (eds.) MICCAI 2020. LNCS, vol. 12263, pp. 364–374. Springer, Cham (2020). https://doi.org/10.1007/978-3-030-59716-0_35
3. Nwoye, C.I., et al.: Rendezvous: attention mechanisms for the recognition of surgical action triplets in endoscopic videos. Med. Image Anal. **78** (2022)
4. Mostafa, M.L., et al.: Surgical flow masked autoencoder for event recognition. Med. Imaging Deep Learn. (2025)
5. Özsoy, E., Örnek, E.P., Eck, U., Czempiel, T., Tombari, F., Navab, N.: 4D-or: semantic scene graphs for or domain modeling. In: International Conference on Medical Image Computing and Computer-Assisted Intervention, pp. 475–485. Springer (2022)
6. Holm, F., Ghazaei, G., Czempiel, T., Özsoy, E., Saur, S., Navab, N.: Dynamic scene graph representation for surgical video. In: Proceedings of the IEEE/CVF International Conference on Computer Vision, pp. 81–87 (2023)
7. Köksal, Ç., Ghazaei, G., Holm, F., Farshad, A., Navab, N.: SANGRIA: surgical video scene graph optimization for surgical workflow prediction. arXiv preprint arXiv:2407.20214 (2024)
8. Saragih, D.G., Hibi, A., Tyrrell, P.N.: Using diffusion models to generate synthetic labeled data for medical image segmentation. Int. J. Comput. Assist. Radiol. Surg. **19**(8), 1615–1625 (2024)
9. Li, C., et al.: Endora: video generation models as endoscopy simulators. In: Linguraru, M.G., et al. (eds.) Medical Image Computing and Computer Assisted Intervention – MICCAI 2024, pp. 230–240, Springer, Heidelberg (2024)
10. Nwoye, C.I.: Surgical text-to-image generation. Pattern Recogn. Lett. (2025)
11. Frisch, Y., et al.: SurGrID: controllable surgical simulation via scene graph to image diffusion. arXiv preprint arXiv:2502.07945 (2025)
12. Yeganeh, Y., et al.: VISAGE: video synthesis using action graphs for surgery. In: Celebi, M.E., Reyes, M., Chen, Z., Li, X. (eds.) Medical Image Computing and Computer Assisted Intervention – MICCAI 2024 Workshops, pp. 146–156, Springer, Cham (2024). https://doi.org/10.1007/978-3-031-77610-6_14
13. Biagini, D., Navab, N., Farshad, A.: HieraSurg: hierarchy-aware diffusion model for surgical video generation. arXiv preprint arXiv:2506.21287 (2025)
14. Rombach, R., Blattmann, A., Lorenz, D., Esser, P., Ommer, B.: High-resolution image synthesis with latent diffusion models. In: Proceedings of the IEEE/CVF Conference on Computer Vision and Pattern Recognition, pp. 10684–10695 (2022)
15. Zhang, L., Rao, A., Agrawala, M.: Adding conditional control to text-to-image diffusion models, pp. 3836–3847 (2023)
16. Gao, S., Lin, Z., Xie, X., Zhou, P., Cheng, M.-M., Yan, S.: EditAnything: empowering unparalleled flexibility in image editing and generation. In: Proceedings of the 31st ACM International Conference on Multimedia, pp. 9414–9416 (2023)
17. Chen, Y., Zhong, K., Wang, F., Wang, H., Zhao, X.: Surgical workflow image generation based on generative adversarial networks. In: 2018 International Conference on Artificial Intelligence and Big Data (ICAIBD), pp. 82–86. IEEE (2018)
18. Marzullo, A., Moccia, S., Catellani, M., Calimeri, F., De Momi, E.: Towards realistic laparoscopic image generation using image-domain translation. Comput. Methods Programs Biomed. **200**, 105834 (2021)

19. Ho, J., Jain, A., Abbeel, P.: Denoising diffusion probabilistic models. In: Advances in Neural Information Processing Systems, vol. 33, pp. 6840–6851 (2020)
20. Lüpke, S., Yeganeh, Y., Adeli, E., Navab, N., Farshad, A.: Physics-informed latent diffusion for multimodal brain MRI synthesis. In: International Conference on Medical Image Computing and Computer-Assisted Intervention, pp. 198–207. Springer (2024)
21. Colleoni, E., Matilla, R.S., Luengo, I., Stoyanov, D.: Guided image generation for improved surgical image segmentation. Med. Image Anal. **97**, 103263 (2024)
22. Hong, S., Lee, J., Woo, S.S.: All but one: surgical concept erasing with model preservation in text-to-image diffusion models. Proc. AAAI Conf. Artif. Intell. **38**, 21143–21151 (2024)
23. Sun, W., Dong, X.-M., Cui, B., Tang, J.: Attentive eraser: unleashing diffusion model's object removal potential via self-attention redirection guidance. Proc. AAAI Conf. Artif. Intell. **39**, 20734–20742 (2025)
24. Radford, A., et al.: Learning transferable visual models from natural language supervision. In: International Conference on Machine Learning, pp. 8748–8763. PMLR (2021)
25. Yuan, K., et al.: Learning multi-modal representations by watching hundreds of surgical video lectures. arXiv preprint arXiv:2307.15220 (2023)
26. Devlin, J., Chang, M.-W., Lee, K., Toutanova, K.: BERT: pre-training of deep bidirectional transformers for language understanding. In: Proceedings of the 2019 Conference of the North American Chapter of the Association for Computational Linguistics: Human Language Technologies, volume 1 (Long and Short Papers), pp. 4171–4186 (2019)
27. Xie, S., Zhang, Z., Lin, Z., Hinz, T., Zhang, K.: SmartBrush: text and shape guided object inpainting with diffusion model. In: Proceedings of the IEEE/CVF Conference on Computer Vision and Pattern Recognition, pp. 22428–22437 (2023)
28. Alabi, O., et al.: CholecinstanceSeg: a tool instance segmentation dataset for laparoscopic surgery. arXiv preprint arXiv:2406.16039 (2024)
29. Ruiz, N., Li, Y., Jampani, V., Pritch, Y., Rubinstein, M., Aberman, K.: DreamBooth: fine tuning text-to-image diffusion models for subject-driven generation. In: Proceedings of CVPR, pp. 22500–22510 (2023)

A Training-Free Approach for 3D Reconstruction from Monocular Sinus Endoscopy

Jan Emily Mangulabnan[1(✉)], Roger D. Soberanis-Mukul[1], Lalithkumar Seenivasan[1], S. Swaroop Vedula[1], Masaru Ishii[2], Gregory Hager[1], Russell H. Taylor[1,2], and Mathias Unberath[1,2]

[1] Johns Hopkins University, Baltimore, MD 21211, USA
{jmangul1,unberath}@jhu.edu
[2] Johns Hopkins Medical Institutions, Baltimore, MD 21287, USA

Abstract. Reliable volumetric representation of the nasal cavity is crucial for enabling quantitative assessment in Functional Endoscopic Sinus Surgery (FESS), yet for most patients the evaluation of their anatomy remains largely qualitative and subjective. While computed tomography (CT) scans can provide 3D anatomical information, their routine use is impractical due to radiation exposure concerns and cost constraints, underscoring the need for a non-invasive alternative. Computer vision methods offer a promising solution for reconstructing sinus anatomy from routine endoscopic video. Current methods rely on Structure-from-Motion (SfM), however, this relies on point correspondences that struggle with photometric inconsistencies inherent to endoscopic imaging, reducing robustness and generalizability. Several sinus reconstruction approaches attempt to mitigate this through learning-based and patient-specific approaches, but suffer from error propagation, leading to inaccurate 3D representations. Optimization-based approaches further introduce excessive training times, limiting their practicality. In this work, we revisit simpler techniques for sinus reconstruction and augment them with track-any-point foundation models to develop a training-free, vision-based 3D reconstruction method. Our approach leverages SfM poses and local point-tracks to generate depth information, recovering a globally consistent structure without fine-tuning requirements. We evaluate our method on six pre-operative endoscopic sequences with respect to the ground-truth CT scan. Our results show that this method improves global geometric accuracy by reducing both point-to-point and pose errors from prior work. Our vision-based approach improves spatial consistency and accuracy in sinus 3D reconstruction, enabling non-invasive postoperative monitoring and seamless clinical integration, offering physicians data-driven insights for improved surgical decision-making.

Keywords: Sinus Endoscopy · 3D Reconstruction · Structure-from-Motion

1 Introduction

Chronic sinusitis, characterized by inflammation of the mucosal lining in the paranasal sinuses, affects one in six adults annually [1,6,17,23], often requiring surgical intervention. Functional Endoscopic Sinus Surgery (FESS) is a widely used treatment that aims to restore normal sinus drainage and ventilation. The longitudinal assessment of the nasal anatomy provides objective and precise measures of postoperative healing and anatomical changes, enabling clinicians to track disease progression, evaluate treatment efficacy, and make data-driven decisions for personalized patient care. However, postoperative management is limited to qualitative assessments based on patient-reported symptoms and direct endoscopic visualization. While computed tomography (CT) scans provide precise structural information, their use for longitudinal assessment is impractical, due to the cost and patient exposure to excessive radiation. Considering the extensive use of endoscopy for postoperative evaluation, vision-based reconstruction of the anatomy directly from endoscopic video sequences offers a promising avenue for tracking anatomical changes over time.

Many reconstruction techniques primarily rely on Structure from Motion (SfM) to determine camera poses and anatomical geometry [3,5,15,16,18,19]. SfM relies on correctly identified point correspondences which are traditionally established by hand-crafted feature descriptors like SIFT. However, the homogeneous appearance of anatomical tissue commonly seen in endoscopic imaging can lead to incorrect matches that propagate and distort the reconstruction. This challenge is further compounded by the inherent photometric variability of endoscopy, where the use of a point light source creates strong specular reflections, non-uniform illumination, and dynamic shading, making feature matching even more difficult to track long-term. While these features often exhibit a limited tracking length, especially in endoscopy, correctly matched features still provide valuable anchor points for geometric consistency [5].

Considering the sparsity of robust feature matches, related works have proposed using hand-crafted descriptors (i.e. SIFT) as a self-supervisory signal in learning-based approaches for reconstruction [9,12,22]. Particularly, Liu et al. [10] leverage SIFT correspondences to guide descriptor learning, resulting in a denser SfM point cloud, even in challenging settings like sinus endoscopy. The generated point cloud then acts as a supervisory signal for monocular depth estimation [8]. When integrated with SfM pose estimation, this allows for patient-specific surface reconstruction through volumetric fusion [2,24]. This method results in visually convincing reconstructions as it enforces self-consistency per sequence. However, it is not necessarily generalizable to new sequences due to its strong reliance on the quality of the self-supervisory signal. Whether due to weak or inconsistent signals, or the intricacies of the optimization process itself, the reconstruction may fail inadvertently for reasons that are difficult to isolate [13].

Recent works have further explored the use of simultaneous localization and mapping (SLAM) in sinus endoscopy [7,21]. In light of the emergence of foundation models (FMs), the track-any-point (TAP)-based method OneS-LAM [21] has shown impressive zero-shot generalization, outperforming domain-specific approaches like EndoSLAM [14] and SAGE-SLAM [7]. While OneSLAM achieves accurate and generalizable local pose estimation, the resulting point cloud remains suboptimal for surface reconstruction when benchmarked against existing methods like [9]. OneSLAM is also limited when applied to longer sequences that capture the sinus cavity beyond the surgical region of interest, where pose estimation begins to exhibit drift and accumulate error. These longer sequences are important for providing a comprehensive anatomical context to support surgical planning, postoperative assessment, and the examination of adjacent or residual pathology outside the immediate area of intervention.

In this work, we reexamine vision-based reconstruction for sinus endoscopy through a new lens, integrating modern tools for computer vision with insights from prior reconstruction efforts. We present a comparably simpler approach that addresses the limitations of current methods described above, leveraging global information from SfM pose estimation to extract structural information with local point tracks from TAP foundation models. Our method takes advantage of the robustness of SIFT feature correspondence for global pose estimation, relying on the observation that while global SfM might generate inaccurate point clouds, its pose estimation is more consistently aligned with the actual camera motion. We use these poses, combined with dense local point correspondence from TAP, to triangulate depth allowing for immediate volumetric fusion via signed distance functions. This results in a generalizable and training-free dense reconstruction method capable of generating surface reconstructions of the sinus anatomy.

2 Methods

Our approach reconstructs 3D sinus anatomy from monocular endoscopic video by leveraging both sparse and dense feature correspondences. Note that all input images are undistorted using intrinsic parameters obtained from a checkerboard-based camera calibration, assuming a pinhole camera model throughout the reconstruction pipeline. First, we estimate camera poses using SIFT features in SfM to establish global consistency. We then recover dense geometry with point tracks from CoTracker [4], a track-any-point foundation model, across a local window of frames. Unlike traditional methods that rely on monocular depth estimation, we directly triangulate the depth of these tracked points with the previously estimated camera poses. The triangulated depth points are backprojected to the 2D image space and integrated into a truncated signed distance function (TSDF) representation to extract the reconstructed surface mesh. An overview of our method is shown in Fig. 1.

2.1 Pose Estimation

We use the COLMAP [20] implementation of SfM with default parameters to create an initial anatomical reconstruction. We leverage SIFT feature correspondences considering their robustness to scale and rotation where these sparse feature correspondences serve as input for bundle adjustment, allowing for the global optimization of camera poses across the sequence. While SIFT is reliable for pose estimation, its short tracking length—due to viewpoint changes, occlusions, and the low-frequency texture of the sinus cavity—limits the number of keypoints that remain consistently tracked across multiple frames. This introduces noise into the sparse point cloud structure, as short track lengths lead to less stable triangulations and increased uncertainty in depth estimation. Therefore, we rely only on SIFT-based SfM to extract camera motion and integrate dense point tracks to accurately recover local structure.

2.2 Dense Depth Triangulation

Using image and camera pose information, we employ a TAP foundation model, CoTracker [4], to establish dense correspondences within a local sequence of k frames. These correspondences, along with calibrated camera poses from SfM, can be used to triangulate dense and accurate local reconstructions using projective geometry. In sinus endoscopy, camera movement mainly occurs along the optical axis which typically produces limited parallax. However, anatomical structures near the periphery can still exhibit sufficient apparent motion. Therefore, we leverage 2D disparity from point tracks in these regions and triangulate depth with the estimated camera poses, bypassing the need for rectification.

To generate local reconstructions, we divide the endoscopic sequence into non-overlapping windows, each with a length of $k = 13$ frames. For each window, we run the TAP model to obtain a set of dense tracks throughout the window. We select points with high-flow information to ensure that the tracks contain enough motion information for structure recovery. Instead of using additional flow estimation models, we rely on geometric approaches for optical flow filtering. First, we identified the pair of frames I_i, I_j within the window with the largest relative camera displacement, c_d. Next, we compute the optical flow between these frames, calculating the magnitude $\Delta \mathbf{x}_{i \to j} = \|\mathbf{x_i} - \mathbf{x_j}\|$ of the optical flow between the set of corresponding points $\mathbf{x_i} \in I_i$, $\mathbf{x_j} \in I_j$ obtained from the TAP model. We then assume that low optical flow associated with the largest camera displacement does not provide significant motion information for structure recovery. Therefore, we remove tracks corresponding to optical flow magnitudes ($\Delta \mathbf{x}_{i \to j}$) below a threshold $\tau_f = 5$ pixels. We assume that valid point tracks will exhibit the most observable motion between these frames due to the increased baseline (c_d). The selected window length ($k = 13$) and flow threshold ($\tau_f = 5$ pixels) parameters were empirically determined to balance spatial coverage and match reliability across sinus sequences. Finally, we use the camera poses and the filtered correspondences within the window to triangulate 3D points for each tracked point, using the direct linear transform (DLT) method. This results in a dense point cloud representing the anatomy section observed in the local temporal window.

Noise can be introduced during the triangulation process because of semi-static points that remain in the sequence and uncertainties in the point-tracking model, which arise from the uniform appearance of the tissue. To minimize the impact of these uncertainties, we apply an additional filter to the point cloud

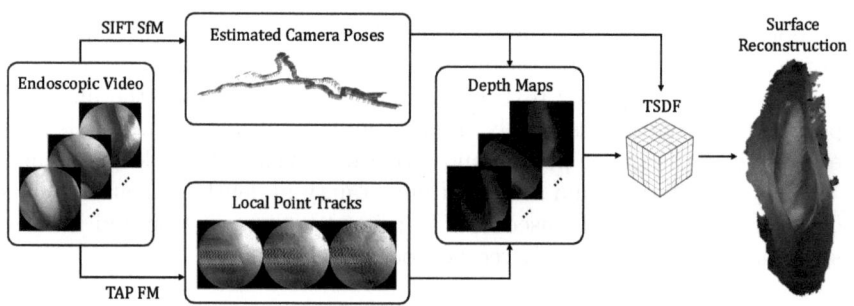

Fig. 1. Overview of Reconstruction Pipeline.

based on estimated depth. The triangulated point cloud is first projected to each image plane to generate a set of sparse depth maps $D_i, i = 0, \ldots, k$. To find the maximum depth value for filtering, we use the minimum optical flow threshold τ_f and the computed camera distance c_d within the window. Considering optical flow as a proxy for disparity we employ τ_f to find the depth associated with this disparity, expressed as: $Z = f \cdot c_d / \tau_f$, where f is focal length. Given c_d and τ_f, any triangulated points that project to a depth larger than the threshold Z are considered geometrically ill-posed, as their corresponding flow would fall below τ_f. These points are therefore discarded to enforce consistency and ensure stable reconstruction. We then apply interpolation to densify the sparse projections in the depth maps. The same process is repeated for all the windows of the sequence to generate depth maps for each camera pose.

2.3 Surface Reconstruction

The dense depth maps obtained from triangulated point tracks are inherently incomplete due to filtering out low-disparity regions, where triangulation confidence is low. This results in sparse or missing regions in the reconstructed depth maps. To resolve these inconsistencies and ensure a more coherent surface representation, we integrate the computed depth estimates into a volumetric representation using TSDF. We employ volumetric fusion [2,24] using the estimated camera poses and depth maps to fuse the information in a common coordinate frame, reinforcing spatial consistency and reducing noise on the reconstructed surfaces. This process effectively fills in gaps, smooths out spurious depth variations, and improves surface completeness by leveraging overlapping depth estimates from different viewpoints. A surface mesh of the anatomy is then extracted using Marching Cubes [11].

3 Experiments and Results

3.1 Experimental Setup

Dataset. We evaluate the reconstruction method on an in-house dataset collected from four human cadaveric subjects. The study describes experiments on ex-vivo specimens that, by definition, do not meet the criterion of human subjects research. Regardless, this study underwent ethical review and was approved as IRB00267324 protocol on non-living human subjects, for which informed consent or ethics agreement was not required. The dataset integrates CT scans from a Brainlab LoopX scanner (Brainlab, Munich, Germany), together with time-paired endoscopic videos recorded with a Storz Image1 HD camera (Karl Storz SE & Co. KG, Tuttlingen, Germany) and camera poses obtained by an NDI Polaris Hybrid Position Sensor (Northern Digital Inc., Waterloo, Canada). The CT scans were preprocessed in 3D Slicer to segment the sinus cavity and generate a ground-truth surface mesh. The optically tracked camera poses were registered to the CT based on segmented anatomical markers mounted on the specimen.

Registration and Evaluation. We reconstruct the sinus cavity from six preoperative endoscopic sequences, leveraging only RGB images as input to the proposed algorithm. The videos consist of 700–1100 frames and view a maximum depth of 40–90 cm, based on depths rendered from the registered poses in the CT segmentation. Our reconstruction spans a broader anatomical region than the surgical site, as surrounding structures can facilitate consistent pairing across time points for longitudinal analysis. Qualitative results of the 3D sinus reconstruction are shown in Fig. 2. Considering that 3D geometry extracted from monocular video has an inherently ambiguous scale and coordinate frame, we scale and register the camera poses estimated from SfM to the corresponding optical tracker poses in order to align the reconstruction in the CT coordinate frame. This scale and transformation is applied to the reconstructed mesh for geometric comparison and validation. We compare our generated reconstructions to the method presented in [9], which evaluated traditional SfM pipelines and demonstrated their failure to produce reconstructions due to sparse and inconsistent geometry. Although OneSLAM [21] is a recent development for pose estimation in the sinus, it still does not outperform [9] in terms of surface mesh quality. As such, [9] remains the state-of-the-art for sinus surface reconstruction for evaluation of our proposed method.

3.2 CT Agreement

Point-to-Mesh Comparison. To evaluate surface-level reconstruction, we compute the point-to-mesh from points sampled on the registered reconstruction to the reference CT segmentation. For each frame in the sequence, we sample a grid of 2D points within the endoscope field of view and project to the reconstructed mesh using depths rendered at the estimated SfM camera

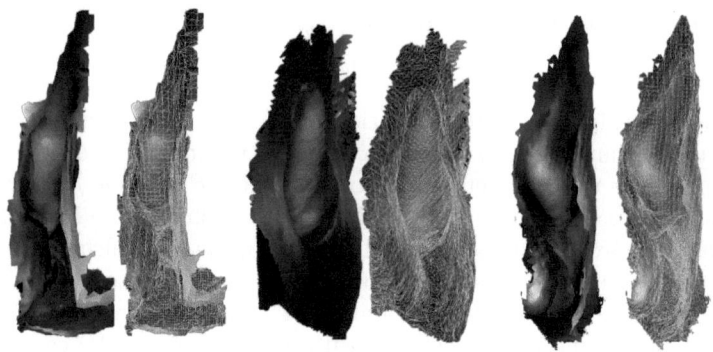

Fig. 2. Qualitative reconstructed mesh results. Note that some surfaces were removed for better visibility.

poses. The resulting set of 3D points is then evaluated by computing the shortest Euclidean distance from each point to the CT surface. This captures how well the reconstructed surface conforms to the anatomical structure, independent of point correspondences or pose alignment. While the point-to-mesh distance offers a measure of overall surface fidelity, we note that this does not guarantee correspondence. Considering the intricate sinus cavity and the extended viewable depth range (40–90 cm), the surface-based comparison may identify nearest points on anatomically incorrect regions, particularly in areas with complex geometry. Therefore, we report point-to-point correspondence errors to directly assess spatial accuracy between the reconstructed and CT anatomy.

Point-to-Point Comparison. We establish approximate 3D point correspondences from the same set of sampled 2D grid points. We leverage the optically-tracked camera poses to project each 2D point using depths rendered from the CT, yielding a set of 3D points on the ground-truth surface. The resulting pair of 3D points from the same 2D point samples are treated as correspondences, to compute the point-to-point error (Euclidean distance) in millimeters. This evaluation is shown in Table 1. considers both structure and viewpoint alignment, offering a more comprehensive evaluation of reconstruction consistency. However, this inherently includes a margin of error introduced by pose estimation uncertainty, so we also report the pose registration error.

Table 1. Quantitative evaluation of generated sinus reconstruction and estimated camera poses with respect to ground-truth CT and optical tracking poses, respectively.

			Pose Registration	
Reconstruction Algorithm	Point-to-Mesh Error (mm)	Point-to-Point Error (mm)	Translation Error (mm)	Rotation Error (°)
Liu et al. [9]	1.62 ± 0.8	6.12 ± 1.7	1.97 ± 1.3	8.13 ± 3.4
Ours	1.15 ± 0.5	5.36 ± 1.9	0.84 ± 0.3	4.71 ± 3.4

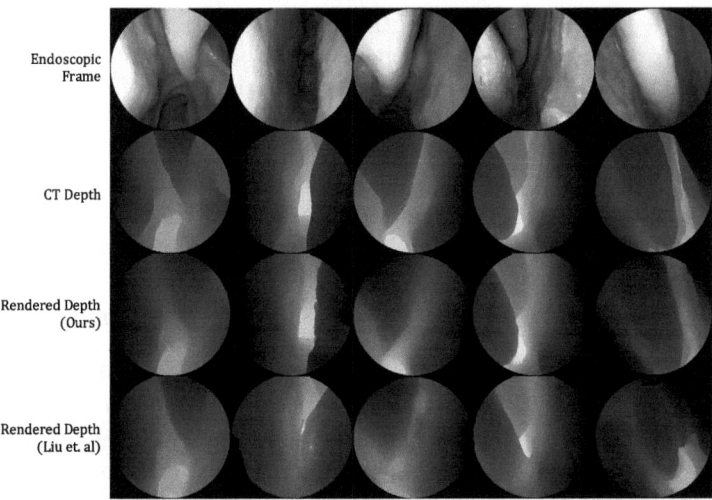

Fig. 3. Input endoscopic frames with depth rendered from CT, our reconstructed meshes, and Liu et al. [9].

4 Discussion and Conclusion

Our results demonstrate that the proposed method outperforms the previous state-of-the-art reconstruction method for sinus endoscopy. These improvements are supported by reduced point-to-mesh and point-to-point accuracy as well as reduced residual errors in the pose registration. The reduced errors suggest that the surface reconstruction more accurately recovers the global geometric structure, as the correspondences more closely align with the anatomy represented by the CT. Since the reconstruction is completely vision-based, both pose and point correspondence also remain consistent considering that the 3D structure is inferred entirely from the camera information.

The improved geometric accuracy ensures that the reconstructed sinus cavity more closely recovers the true anatomy. Compared to monocular depth estimation methods, this vision-based approach provides a more consistent and interpretable reconstruction, as it does not rely on learned priors but rather enforces multi-view geometric constraints. By computing depth directly from camera poses and point correspondences, our method eliminates the need for fine-tuning and offers a more predictable and geometrically constrained reconstruction compared to deep learning models, preventing scale drift and ensuring more robust spatial alignment. As shown in Fig. 3, our reconstruction more closely matches the CT depth rendering, demonstrating improved spatial consistency compared to the previous method, which exhibits greater shape distortions due to per-frame depth estimation errors.

While SfM can be sensitive to low-texture and specular reflection present in endoscopy, it produced stable and accurate trajectories in the evaluated sequences. All frames were successfully localized, yielding low residual errors when registered to ground-truth optically tracked poses. Despite the limited number of sequences, the method demonstrated both qualitative and quantitative alignment with CT and tracking data, supporting its feasibility as a practical alternative for image-based sinus reconstruction. Future work will extend evaluation to a broader range of sequences capturing greater anatomical variability and potential surgical artifacts to improve robustness across patient anatomy and assess applicability for intraoperative use.

Our pipeline offers a more streamlined approach compared to the benchmark method [9] which requires dense descriptors and depth model fine-tuning. Both pipelines require SfM, but our method avoids the dense descriptor computation and additional training, and TAP inference is considerably simpler than the patient-specific fine-tuning required for dense depth estimation in the benchmark pipeline. As SfM and TAP still incur some computational overhead, further optimization is likely necessary for clinical integration, but the modular, training-free nature of our approach is a promising step toward real-world applicability.

By leveraging only vision-based constraints, our approach enhances spatial consistency, interpretability, and adaptability for surgical applications, offering a viable alternative to learned depth models. This method has the potential to support broader clinical applications, including longitudinal and quantitative assessments for postoperative monitoring. Future work can further explore real-time surgical integration with the potential to bridge into intraoperative model updating, enabling dynamic and continuously refined 3D reconstructions to support more adaptive and responsive surgical navigation.

References

1. Bhattacharyya, N.: Incremental health care utilization and expenditures for chronic rhinosinusitis in the United States. Ann. Otol. Rhinol. Laryngolo. **120**, 423–427 (2011). https://doi.org/10.1177/000348941112000701
2. Curless, B., Levoy, M.: A volumetric method for building complex models from range images. In: Proceedings of the 23rd Annual Conference on Computer Graphics and Interactive Techniques (1996). https://api.semanticscholar.org/CorpusID:12358833
3. Hernández, I., et al.: Investigating keypoint descriptors for camera relocalization in endoscopy surgery. Int. J. Comput. Assisted Radiol. Surge. **18** (2023). https://doi.org/10.1007/s11548-023-02918-x
4. Karaev, N., Makarov, I., Wang, J., Neverova, N., Vedaldi, A., Rupprecht, C.: CoTracker3: simpler and better point tracking by pseudo-labelling real videos. arXiv preprint arXiv:2410.11831 (2024)
5. Leonard, S., et al.: Evaluation and stability analysis of video-based navigation system for functional endoscopic sinus surgery on in vivo clinical data. IEEE Trans. Med. Imaging **37**(10), 2185–2195 (2018). https://doi.org/10.1109/TMI.2018.2833868
6. Lethbridge-Cejku, M., Schiller, J., Bernadel, L.: Summary health statistics for U.S. adults: national health interview survey, 2002. Vital Health Stat. Ser. 10 Data from Nat. Health Surv. **10**, 1–151 (08 2004)

7. Liu, X., Li, Z., Ishii, M., Hager, G.D., Taylor, R.H., Unberath, M.: SAGE: slam with appearance and geometry prior for endoscopy. In: 2022 International Conference on Robotics and Automation (ICRA), pp. 5587–5593. IEEE (2022)
8. Liu, X., et al.: Dense depth estimation in monocular endoscopy with self-supervised learning methods. IEEE Trans. Med. Imaging **39**(5), 1438–1447 (2020). https://doi.org/10.1109/TMI.2019.2950936
9. Liu, X., et al.: Reconstructing sinus anatomy from endoscopic video - towards a radiation-free approach for quantitative longitudinal assessment. In: Martel, A.L., et al. (eds.) Medical Image Computing and Computer Assisted Intervention - MICCAI 2020, pp. 3–13. Springer International Publishing, Cham (2020)
10. Liu, X., et al.: Extremely dense point correspondences using a learned feature descriptor. In: Proceedings of the IEEE/CVF Conference on Computer Vision and Pattern Recognition, pp. 4847–4856 (2020)
11. Lorensen, W., Cline, H.: Marching cubes: a high resolution 3D surface construction algorithm. ACM SIGGRAPH Comput. Graph. **21**, 163 (1987). https://doi.org/10.1145/37401.37422
12. Ma, R., Wang, R., Pizer, S., Rosenman, J., McGill, S.K., Frahm, J.-M.: Real-time 3D reconstruction of colonoscopic surfaces for determining missing regions. In: Shen, D., et al. (eds.) MICCAI 2019. LNCS, vol. 11768, pp. 573–582. Springer, Cham (2019). https://doi.org/10.1007/978-3-030-32254-0_64
13. Mangulabnan, J.E., et al.: A quantitative evaluation of dense 3D reconstruction of sinus anatomy from monocular endoscopic video. arXiv preprint arXiv:2310.14364 (2023)
14. Ozyoruk, K.B., et al.: EndoSLAM dataset and an unsupervised monocular visual odometry and depth estimation approach for endoscopic videos. Med. Image Anal. **71**, 102058 (2021)
15. Phan, T.B., Trinh, H., Lamarque, D., Wolf, D., Daul, C.: Dense optical flow for the reconstruction of weakly textured and structured surfaces: application to endoscopy, pp. 310–314 (2019). https://doi.org/10.1109/ICIP.2019.8802948
16. Qiu, L.: Endoscope navigation and 3D reconstruction of oral cavity by visual slam with mitigated data scarcity, pp. 2278–22787 (2018). https://doi.org/10.1109/CVPRW.2018.00295
17. Ray, N.F., et al.: Healthcare expenditures for sinusitis in 1996: contributions of asthma, rhinitis, and other airway disorders. J. Allergy Clin. Immunol. **103 3 Pt 1**, 408–414 (1999). https://api.semanticscholar.org/CorpusID:26049225
18. Reiter, A., Léonard, S., Sinha, A., Ishii, M., Taylor, R.H., Hager, G.: Endoscopic-CT: learning-based photometric reconstruction for endoscopic sinus surgery. SPIE Med. Imaging (2016). https://api.semanticscholar.org/CorpusID:22169509
19. Resindra, A., Monno, Y., Okutomi, M., Suzuki, S., Gotoda, T., Miki, K.: Whole stomach 3D reconstruction and frame localization from monocular endoscope video. IEEE J. Transl. Eng. Health Med. **7**, 1 (2019). https://doi.org/10.1109/JTEHM.2019.2946802
20. Schonberger, J.L., Frahm, J.M.: Structure-from-motion revisited. In: Proceedings of the IEEE Conference on Computer Vision and Pattern Recognition, pp. 4104–4113 (2016)
21. Teufel, T., et al.: OneSLAM to map them all: a generalized approach to SLAM for monocular endoscopic imaging based on tracking any point. Int. J. Comput. Assist. Radiol. Surg., 1–8 (2024)
22. Turan, M., Pilavci, Y.Y., Ganiyusufoglu, I., Araujo, H., Konukoglu, E., Sitti, M.: Sparse-then-dense alignment-based 3D map reconstruction method for endoscopic

capsule robots. Mach. Vis. Appl. **29**(2), 345–359 (2017). https://doi.org/10.1007/s00138-017-0905-8
23. Wyler, B., Mallon, W.: Sinusitis update. Emerg. Med. Clin. North Am. **37**, 41–54 (2019). https://doi.org/10.1016/j.emc.2018.09.007
24. Zach, C., Pock, T., Bischof, H.: A globally optimal algorithm for robust TV-L1 range image integration, pp. 1–8 (2007). https://doi.org/10.1109/ICCV.2007.4408983

TrackOR: Towards Personalized Intelligent Operating Rooms Through Robust Tracking

Tony Danjun Wang[1]($^{\boxtimes}$), Christian Heiliger[3], Nassir Navab[1,2], and Lennart Bastian[1,2]

[1] Computer Aided Medical Procedures, TU Munich, Munich, Germany
tony.wang@tum.de
[2] Munich Center for Machine Learning, Munich, Germany
[3] Minimally Invasive Surgery, University Hospital of Munich (LMU), Munich, Germany

Abstract. Providing intelligent support to surgical teams is a key frontier in automated surgical scene understanding, with the long-term goal of improving patient outcomes. Developing personalized intelligence for all staff members requires maintaining a consistent state of *who* is located *where* for long surgical procedures, which still poses numerous computational challenges. We propose *TrackOR*, a framework for tackling long-term multi-person tracking and re-identification in the operating room. *TrackOR* uses 3D geometric signatures to achieve state-of-the-art online tracking performance (+11% Association Accuracy over the strongest baseline), while also enabling an effective offline recovery process to create analysis-ready trajectories. Our work shows that by leveraging 3D geometric information, persistent identity tracking becomes attainable, enabling a critical shift towards the more granular, staff-centric analyses required for personalized intelligent systems in the operating room. This new capability opens up various applications, including our proposed *temporal pathway imprints* that translate raw tracking data into actionable insights for improving team efficiency and safety and ultimately providing personalized support.

Keywords: Surgical Data Science · Multiple-Object Tracking · Person Re-Identification

1 Introduction

The operating room (OR) is the quintessential high-stakes environment – small actions can lead to profound consequences. To improve patient outcomes, Surgical Data Science (SDS) has emerged with the goal of creating a data-driven feedback loop to make the OR safer and more efficient [20]. Historically, SDS has centered on the surgeon, analyzing instrument handling and technique, often

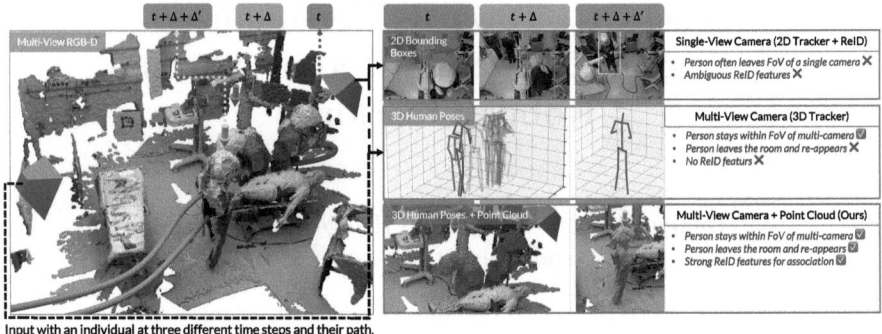

Fig. 1. We introduce *TrackOR*, a framework for long-term multi-person tracking in the OR. Unlike conventional 2D trackers that struggle with field-of-view (FoV) limits and appearance-based ReID, or standard 3D trackers that lack features for re-identification after prolonged absences, *TrackOR* uses 3D geometric signatures to maintain persistent identity even when staff leave and re-enter the room.

through the viewpoint of endoscopes or laparoscopes [10]. Recognizing that surgery is a complex team endeavor, the field's focus has recently broadened to encompass the entire surgical team's dynamics and collaborative patterns [29]. Yet this wider perspective still operates at a coarse granularity, typically analyzing workflow at the role level, treating, for instance, the "circulating nurse" as an archetype rather than an individual [21,31].

We posit that the next generation of surgical domain models will be driven by personalized intelligent systems, which in turn requires a fundamental shift from a role-based to a *staff-centric* understanding of the OR [29]. This shift moves beyond generic roles by recognizing that each staff member has unique skill levels and develops distinct habits over time, e.g., when adapting to new surgical instruments or team members. However, this leap towards personalized intelligence requires new capabilities: generating persistent, long-term trajectories for each staff member, even across extended absences from the OR.

Multi-object tracking (MOT) is a notoriously difficult task with numerous challenges. Strikingly, the OR represents a confluence of many of these challenges, including severe occlusions and overall crowdedness [2]. These challenges are exacerbated by staff wearing visually indistinct homogeneous attire, which has rendered prior OR tracking methods fundamentally identity-agnostic. While these methods can handle the simple, short-term task of associating an individual from one frame to the next, they inevitably fail when faced with the 'revolving door[1]' reality of surgical procedures, confining their utility to uninterrupted video segments and making longitudinal analysis challenging.

To overcome this, a robust global re-identification (ReID) capability is not just beneficial but essential. While many tracking methods incorporate appearance-based ReID [9], such methods are bound to fail when confronted

---

[1] Refers to, e.g., the circulating nurse shuttling between ORs.

with the visual homogeneity of the OR [29]. A more powerful paradigm is needed, where identity is derived from a robust, view-invariant signature that is decoupled from confounding visual and textural cues. The effectiveness of such a paradigm, however, depends largely on the choice of its underlying data representation, for example, 2D appearance, 3D pose, or 3D geometry (see Fig. 1). In this paper, we investigate these options and propose a framework centered on 3D geometric signatures as a robust solution. Our approach leverages point cloud representations for online association, also enabling an offline process to correct tracking errors and merge fragmented tracklets, reconstructing complete, persistent journeys of each staff member. Our **contributions** are summarized as follows:

- We propose *TrackOR*, a novel end-to-end tracking framework that overcomes the limitations of appearance-based methods by integrating a robust, view-invariant ReID signature for persistent, long-term identity tracking.
- We introduce a method that combines 3D pose and ReID for online tracking and leverages ReID alone for offline global trajectory recovery, showing a unique synergy for solving the tracking problem in the OR.
- We propose *temporal pathway imprints* and demonstrate how our framework unlocks long-term workflow and pathway analysis with these imprints.

2 Related Works

Multi-Object Tracking (MOT) represents a formidable challenge in the operating room (OR). While standard trackers typically combine a motion model, like the Kalman filter [13], for short-term prediction [8,30] with an appearance based Re-Identification (ReID) module to handle longer occlusions [1,11,19,26], these approaches are confounded by the complex dynamics and visually homogeneous OR environment. Consequently, prior OR-specific approaches, whether they reconstruct and associate in 3D directly ("3D-first") [5,21] or track in 2D before aggregating ("2D-first") [12], have largely been identity-agnostic, limiting their utility to short-term analysis. While recent work has shown the promise of non-texture-based signatures for the isolated ReID task [29], a complete framework that effectively integrates this principle for robust, long-term tracking has remained elusive. Our work bridges the gap between the isolated Re-Identification and robust end-to-end tracking; this comprehensive framework leverages non-texture-based signatures to recover the global, long-term trajectories of OR staff.

3 Unifying Short- and Long-Term Tracking

Given a video sequence of T frames, each frame t contains a set of detections $D_t = \{d_t^1, d_t^2, \ldots, d_t^{N_t}\}$. Our objective is to solve a multi-object tracking problem with long-term re-identification. The goal is to partition the detections from all

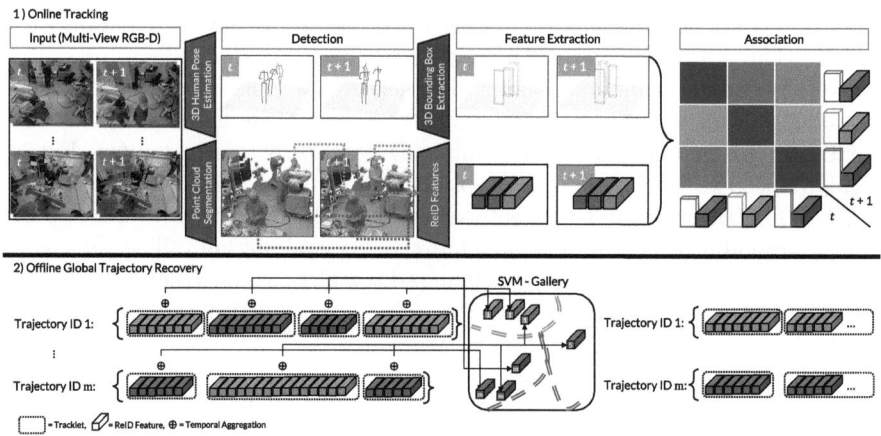

Fig. 2. Overview of our online (top) and offline trajectory recovery (bottom). Online, we use multi-view RGB-D frames to extract spatial and ReID features from 3D poses and point clouds, creating a cost matrix for data association. Offline, we classify the resulting tracklets using an SVM-Gallery [29] and reconstruct each person's global trajectory based on this classification.

frames into a set of trajectories $\mathcal{T} = \{\tau_1, \tau_2, ..., \tau_M\}$, where M is the total number of unique individuals observed.

In the OR, however, the tracking problem is compounded by staff frequently leaving and re-entering the scene for extended periods. These long-term absences cause simple, continuous trajectory models to be ineffective.

We first define a *tracklet* trk_m^a as a single, continuous period of visibility for an individual m. It is formally defined as a time-ordered sequence of states:

$$\text{trk}_m^a = \{s_t^m\}_{t=b_m^a}^{e_m^a},$$

where s_t^m is the state of individual m at frame t (e.g., a bounding box), and the indices b_m^a and e_m^a denote the beginning and ending frames of that specific appearance. Building on this, we define a *trajectory* τ_m as the complete history of an individual, represented by a collection of their tracklets:

$$\tau_m = \{\text{trk}_m^1, \text{trk}_m^2, ..., \text{trk}_m^{A_m}\},$$

where A_m is the number of distinct appearances for individual m. This formulation explicitly models the periods of absence between tracklets, making it well-suited for the tracking problem in the OR. This formulation encompasses both short-term and long-term absences between different surgical procedures.

3.1 TrackOR

In Fig. 2, we illustrate the overall pipeline of our proposed method, TrackOR. TrackOR follows the "tracking by detection" paradigm [9], consisting of detection, feature extraction, and association (Fig. 2 top). Moreover, with our framework, we can further leverage extracted ReID features to perform a straightforward and effective offline global trajectory recovery (Fig. 2 bottom).

Detection. Following prior '3D-first' approaches [5,16,21], we directly detect 3D human poses from the multi-view RGB data. Specifically, we adopt VoxelPose [27], which aggregates heatmap predictions from each 2D view into a unified 3D voxel volume. This volume is then processed by a 3D CNN that predicts the root location of each person. For each predicted root location, another 3D CNN then regresses the final, detailed 3D poses. For a complete description of this process, we refer the reader to the original publication [27].

Feature Extraction. We obtain the person ReID features $\mathcal{F}_t^i$ from the segmented 3D point cloud of each 3D human pose detection. Specifically, we first segment the 3D point cloud of the entire scene to obtain 3D point clouds of each human and associate each 3D human pose with their respective 3D human point cloud [3]. Subsequently, each 3D human point cloud is projected into 8 virtual camera viewpoints, positioned around each object in an equidistant circular arrangement, to generate a set of 2D depth maps that are further processed with a ReID network [29] to obtain our final ReID feature vectors $\mathcal{F}_t^i \in \mathbb{R}^{8 \times C}$, where C is the feature dimension.

Association. At this point, for each frame t, we have a set of 3D human pose detections D_t and a set of active trajectories $\mathcal{T}_{t-1}$ from the previous frame. Each detection d_t^i and trajectory τ_m is represented by a ReID feature vector and a 3D bounding box. To associate new detections with existing trajectories, we compute a cost matrix **C** where each entry represents the dissimilarity between a detection and a trajectory. We define the cost as a weighted sum of the shape cost, based on the cosine dissimilarity of the ReID features, and a spatial cost, based on the 3D Generalized IoU (GIoU) [22] between bounding boxes. With the cost matrix defined, we employ a linear assignment strategy using the Hungarian algorithm [14] to find the optimal matching. Associations with a final cost higher than a predefined threshold γ are discarded. Matched detections update their respective trajectory states, unmatched detections initialize new tracks, and unmatched trajectories are marked as "lost".

Global Trajectory Recovery. For downstream tasks, we perform a final offline recovery step to correct fragmentation and identity switches from the online stage (Fig. 2 bottom). To do this, we first apply temporal max-pooling to aggregate each tracklet's sequence of feature vectors $\mathbb{R}^{l \times 8 \times C}$ into a single representative descriptor per view. An SVM-Gallery [15,29] then assigns an identity to this descriptor via a majority vote over its 8 view-specific feature vectors. Finally, all tracklets assigned the same identity are grouped to reconstruct the complete, global trajectory for each person.

Temporal Pathway Imprint. To extract insights from the obtained global trajectories, we propose *temporal pathway imprints*. We achieve this by projecting the root positions onto the X-Y plane of the OR and viewing the result from

Table 1. Quantitative comparison of TrackOR (Ours) against 2D and 3D tracking baselines on the MM-OR test set. Bold indicates the best performance per metric. ↑ Higher is better, ↓ lower is better, † denotes offline methods.

Model		HOTA$_{(\alpha=.05-.5)}$ [17]			Identity [23]		CLEAR [7]			Count		Speed
Tracker	ReID	HOTA↑	AssA↑	DetA↑	IDF1↑	IDSW↓	MOTA↑	FP↓	FN↓	% #Dets	% #IDs	FPS↑
2D Bounding Box Tracker, using [28] as detections												
OC-Sort [8]	✗	49.660	27.054	91.158	40.089	566	79.575	849	**566**	102.92	362.26	850
ByteTrack [30]	✗	58.430	37.451	91.163	52.705	312	75.946	950	1071	98.75	141.50	997
Strong Sort [11]	RGB	57.965	36.783	91.347	43.378	377	72.729	679	744	**99.33**	164.15	20
Boost Track [26]	RGB	54.848	33.013	91.130	42.989	511	77.420	622	1057	95.52	1,037.74	60
Deep OC-Sort [19]	RGB	78.359	66.348	**92.545**	73.007	200	**85.256**	437	793	97.36	239.62	34
BoT Sort [1]	RGB	80.825	71.309	91.612	74.686	266	78.936	837	940	98.94	139.62	32
3D Human Pose Tracker, using [24] as detections												
KSP Tracker† [5]	✗	54.037	36.086	80.918	46.369	462	51.768	1566	2650	88.82	158.49	115
Nearest-Neighbor [21]	✗	73.366	65.813	81.787	66.496	86	55.686	1564	2648	88.82	113.21	2365
Kalman Filter [16]	✗	71.047	62.441	80.840	63.317	**80**	54.686	1543	2772	88.82	**101.88**	1121
TrackOR (Ours)	Depth	**82.216**	**82.300**	83.685	**76.362**	125	55.284	1564	2648	88.82	130.19	17

a bird's-eye view. Adding regions of interest, such as sterility zones, can further enhance the context. Analagous to Wang et al.'s [29] *3D activity imprints* which depict the duration of time at each spatial location, we model an individual's temporal pathway through the OR as a function of time.

4 Experiments and Results

Datasets. Due to privacy concerns, publicly available datasets of surgical procedures captured from ceiling-mounted cameras are generally scarce [4]. While previous datasets such as MVOR [25] and 4D-OR [21] exist, they are not suitable for developing robust, long-term tracking solutions. MVOR [25] lacks the necessary temporal continuity and identity annotations, while 4D-OR [21] is overly simplified due to minimal movement and obstructions, allowing near-perfect tracking with a simple nearest-neighbor association [29]. As such, we perform our experiments on the recently introduced MM-OR dataset [31], which, in contrast, exhibits tracking challenges, including frequent occlusions, multiple clinicians working in tight spaces, and homogenous attire. For our experiments, we use the multi-view RGB-D data from the ceiling-mounted cameras, which provide three annotated RGB views and a corresponding 3D point cloud for each frame. In our study, we use all takes in MM-OR that provide segmentation labels. From the resulting 20 takes (totaling 23,442 frames), we partition 13 takes (62%) into a train set, 2 takes (16%) into a validation set, and 5 takes (22%) into a test set. In MM-OR, each take corresponds to a single surgery. We consulted the dataset's authors to obtain identity mappings.

Metrics. To evaluate the broad range of tracking methods (2D and 3D) consistently, we assess all methods in the 2D image space. Accordingly, maintaining consistent IDs across different camera views is not required by our evaluation protocol. For 3D methods, we project 3D pose detections into each 2D image plane to generate bounding boxes. We follow standard multi-object tracking benchmarks and report the Higher Order Tracking Accuracy (HOTA) [17] with its sub-components, as well as the classic CLEAR [7], Identity [23], and Counting metrics. To account for the inherent geometric discrepancies between bounding boxes derived from 3D poses and the 2D ground truth derived from silhouettes, we adjust the HOTA α-range to 0.05–0.5.

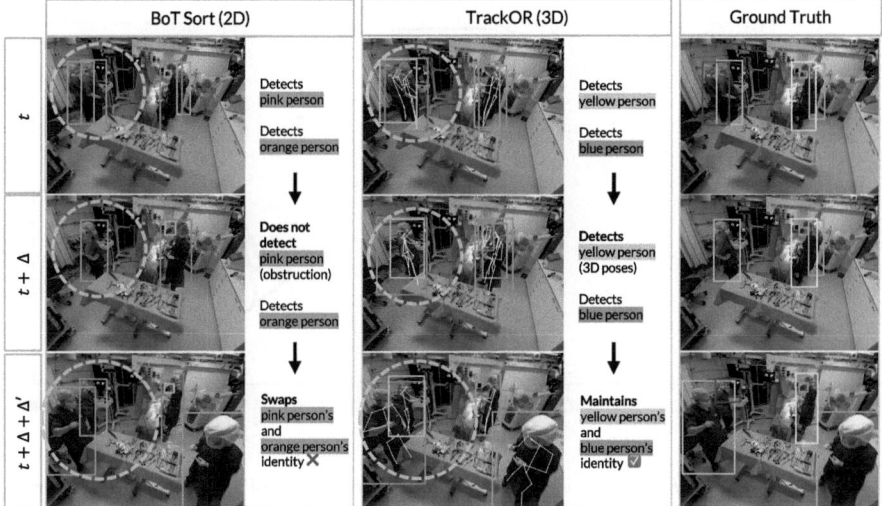

Fig. 3. Qualitative results of BoT-Sort [1] and TrackOR (Ours). The bounding box colors reflect the predicted identity.

4.1 Implementation Details

Detection Backbones. To ensure a fair comparison, all evaluated methods use detections from the same backbone. For 2D tracking baselines, we provide detections from a YOLOv10-B [28] model fine-tuned on our training set. For all 3D-based methods (including ours), we use 3D human pose estimations as detection. Since MM-OR lacks 3D annotations, we train a pose estimation network using the self-supervised approach proposed in [24]. 3D point cloud segmentations are obtained by projecting ground truth segmentation masks into 3D.

Re-identification Modules. The appearance-based 2D trackers utilize a standard feature extractor [18] fine-tuned on cropped images of individuals from our training split. In contrast, our method's 3D ReID module uses a ResNet-9 backbone [29] to extract view-invariant features from person-specific point clouds, likewise trained on the identity labels of our training set.

Evaluated Trackers. We compare our framework against a comprehensive set of baselines. The 2D trackers include OCSort [8] and ByteTrack [30] (without ReID), as well as the ReID-based methods BoostTrack++ [26], DeepOCSort [19], StrongSort [11], and BoTSort [1]. Our 3D baselines implement the tracking paradigms proposed by Belagiannis et al. [5,6], Ozsoy et al. [21], and Liu et al. [16]. For all methods, tracking hyperparameters are tuned on the validation set.

4.2 Results

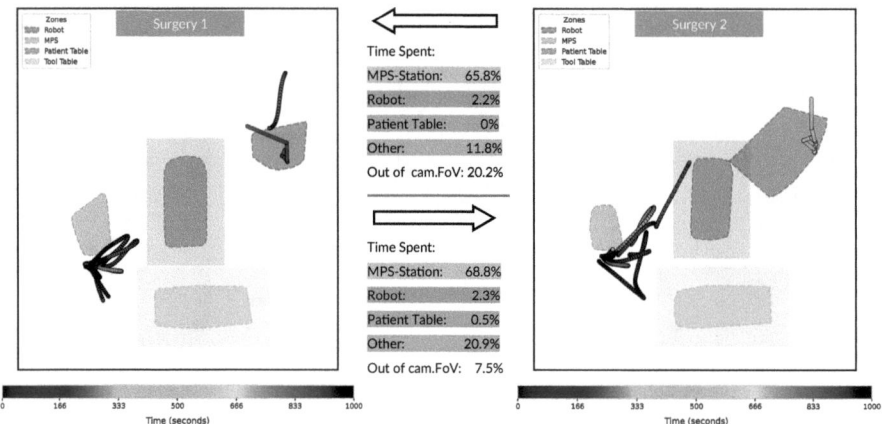

Fig. 4. *Temporal pathway imprints* of the robot technician of two different surgeries. We plot the first 1,000 s of each surgery. The extended transparent borders delineate the 12-inch border around the sterile fields.

Tracking Performance. The quantitative results in Table 1 show that the online results of TrackOR (Ours) score the highest overall HOTA [17] of 82.2%. This performance is driven by the Association Accuracy (AssA) of 82.3%, which is over 11% higher than the best baseline (BoT Sort [1]). This result confirms the superiority of our geometric ReID feature for maintaining correct identities through the challenging occlusions and visual homogeneity of the OR. Conversely, the results reveal that all 3D-based methods exhibit a lower detection accuracy (DetA) compared to the 2D baselines. This is an expected trade-off, as our 3D pose estimation backbone is trained using a self-supervised framework, compared to the 2D detector fine-tuned on ground-truth bounding boxes. Consequently, older metrics that are heavily biased towards detection, like MOTA [7], naturally favor the 2D approaches. This trade-off also extends to computational cost, as processing 3D data results in a lower framerate (FPS) for TrackOR compared to the faster, single-modality 2D trackers (measured on a single RTX 2080 Ti).

Figure 3 illustrates qualitative results of BoT Sort (with ReID) [1] and TrackOR. The sequence highlights a common failure mode where 2D trackers lose a person during occlusion and subsequently switch their ID due to visual similarity. Our method, operating in 3D, successfully navigates both the occlusion and the re-identification challenge, maintaining the correct identities throughout.

Downstream Task: Temporal Pathway Imprints. As a downstream application of our tracker, Fig. 4 shows two *temporal pathway imprints* from the same robot technician across two different surgeries. The imprints reveal that while the technician's primary workspace was the MPS station in both procedures, they interacted with the robot twice in Surgery 1 compared to only once in Surgery 2. More critically, the pathway in Surgery 2 captures the non-sterile technician coming into close proximity with the sterile patient table. Ultimately, these imprints demonstrate the potential to move towards a data-driven science of the OR, enabling automated workflow analysis, objective safety monitoring, and personalized feedback for the entire surgical team.

5 Conclusion

We presented TrackOR, a novel framework for robust, long-term multi-person tracking and re-identification in the operating room. Prior methods often fail to maintain identity through the prolonged absences of staff common in surgery, a challenge exacerbated by visually similar attire that confounds appearance-based ReID. Our approach addresses these challenges by deriving identity from robust 3D geometric signatures, which enables fine-grained, person-centric workflow analyses, as demonstrated by our *temporal pathway imprints*. We will make all codes public as we believe this work is an important step towards the next generation of personalized intelligent systems capable of providing tailored support to the entire surgical team.

Acknowledgments. This work was partly supported by the state of Bavaria through Bayerische Forschungsstiftung (BFS) under Grant AZ-1592-23-ForNeRo and the German Federal Ministry for Economic Affairs and Climate Action (BMWK) through the Central Innovation Programme for SMEs (ZIM) under Grant KK 5389102BA3.

Disclosure of Interests. The authors have no competing interests to declare that are relevant to the content of this article.

References

1. Aharon, N., Orfaig, R., Bobrovsky, B.Z.: BoT-SORT: robust associations multi-pedestrian tracking. arXiv preprint arXiv:2206.14651 (2022)
2. Bastian, L., et al.: Know your sensors—a modality study for surgical action classification. In: MICCAI 2022 AE-CAI Workshop, vol. 11, no. 4, pp. 1113–1121 (2023)
3. Bastian, L., Derkacz-Bogner, D., Wang, T.D., Busam, B., Navab, N.: SegmentOR: obtaining efficient operating room semantics through temporal propagation. In: MICCAI, pp. 57–67. Springer (2023)
4. Bastian, L., Wang, T.D., Czempiel, T., Busam, B., Navab, N.: DisguisOR: holistic face anonymization for the operating room. IJCARS, 1–7 (2023)

5. Belagiannis, V., et al.: Parsing human skeletons in an operating room. Mach. Vis. Appl. **27**(7), 1035–1046 (2016). https://doi.org/10.1007/s00138-016-0792-4
6. Berclaz, J., Fleuret, F., Turetken, E., Fua, P.: Multiple object tracking using K-shortest paths optimization. IEEE Trans. Pattern Anal. Mach. Intell. **33**(9), 1806–1819 (2011)
7. Bernardin, K., Stiefelhagen, R.: Evaluating multiple object tracking performance: the clear mot metrics. EURASIP J. Image Video Process. **2008**, 1–10 (2008)
8. Cao, J., Pang, J., Weng, X., Khirodkar, R., Kitani, K.: Observation-centric sort: rethinking sort for robust multi-object tracking. In: Proceedings of the IEEE/CVF Conference on Computer Vision and Pattern Recognition, pp. 9686–9696 (2023)
9. Ciaparrone, G., Sánchez, F.L., Tabik, S., Troiano, L., Tagliaferri, R., Herrera, F.: Deep learning in video multi-object tracking: a survey. Neurocomputing **381**, 61–88 (2020)
10. Czempiel, T., Sharghi, A., Paschali, M., Navab, N., Mohareri, O.: Surgical workflow recognition: from analysis of challenges to architectural study. In: ECCV Workshops. Springer (2022)
11. Du, Y., et al.: StrongSORT: make DeepSORT great again. IEEE Trans. Multimedia **25**, 8725–8737 (2023)
12. Hu, H., Hachiuma, R., Saito, H., Takatsume, Y., Kajita, H.: Multi-camera multi-person tracking and re-identification in an operating room. J. Imaging **8**(8) (2022)
13. Kalman, R.E.: A new approach to linear filtering and prediction problems. J. Basic Eng. **82**(1), 35–45 (1960)
14. Kuhn, H.W.: The Hungarian method for the assignment problem. Naval Res. Logistics Q. **2**(1–2), 83–97 (1955)
15. Layne, R., et al.: A dataset for persistent multi-target multi-camera tracking in RGB-D. In: Proceedings of the IEEE Conference on Computer Vision and Pattern Recognition Workshops, pp. 47–55 (2017)
16. Liu, B., et al.: A human mesh-centered approach to action recognition in the operating room. Artif. Intell. Surg. **4**(2), 92–108 (2024)
17. Luiten, J., et al.: HOTA: a higher order metric for evaluating multi-object tracking. Int. J. Comput. Vision **129**, 548–578 (2021)
18. Luo, H., Gu, Y., Liao, X., Lai, S., Jiang, W.: Bag of tricks and a strong baseline for deep person re-identification. In: Proceedings of the IEEE/CVF Conference on Computer Vision and Pattern Recognition Workshops (2019)
19. Maggiolino, G., Ahmad, A., Cao, J., Kitani, K.: Deep OC-SORT: multi-pedestrian tracking by adaptive re-identification. In: 2023 IEEE International Conference on Image Processing (ICIP), pp. 3025–3029. IEEE (2023)
20. Maier-Hein, L., et al.: Surgical data science-from concepts toward clinical translation. Med. Image Anal. **76**, 102306 (2022)
21. Özsoy, E., Örnek, E.P., Eck, U., Czempiel, T., Tombari, F., Navab, N.: 4D-OR: semantic scene graphs for or domain modeling. In: MICCAI. Springer (2022)
22. Rezatofighi, H., Tsoi, N., Gwak, J., Sadeghian, A., Reid, I., Savarese, S.: Generalized intersection over union. In: The IEEE Conference on Computer Vision and Pattern Recognition (CVPR), June 2019
23. Ristani, E., Solera, F., Zou, R., Cucchiara, R., Tomasi, C.: Performance measures and a data set for multi-target, multi-camera tracking. In: Hua, G., Jégou, H. (eds.) ECCV 2016. LNCS, vol. 9914, pp. 17–35. Springer, Cham (2016). https://doi.org/10.1007/978-3-319-48881-3_2
24. Srivastav, V., Chen, K., Padoy, N.: SelfPose3D: self-supervised multi-person multi-view 3D pose estimation. In: Proceedings of the IEEE/CVF Conference on Computer Vision and Pattern Recognition (CVPR), pp. 2502–2512, June 2024

25. Srivastav, V., Issenhuth, T., Abdolrahim, K., de Mathelin, M., Gangi, A., Padoy, N.: MVOR: a multi-view RGB-D operating room dataset for 2D and 3D human pose estimation (2018)
26. Stanojević, V., Todorović, B.: BoostTrack++: using tracklet information to detect more objects in multiple object tracking. arXiv preprint arXiv:2408.13003 (2024)
27. Tu, H., Wang, C., Zeng, W.: VoxelPose: towards multi-camera 3D human pose estimation in wild environment. In: Vedaldi, A., Bischof, H., Brox, T., Frahm, J.-M. (eds.) ECCV 2020. LNCS, vol. 12346, pp. 197–212. Springer, Cham (2020). https://doi.org/10.1007/978-3-030-58452-8_12
28. Wang, A., et al.: YOLOv10: real-time end-to-end object detection. In: Advances in Neural Information Processing Systems, vol. 37, pp. 107984–108011 (2024)
29. Wang, T.D., Bastian, L., Czempiel, T., Heiliger, C., Navab, N.: Beyond role-based surgical domain modeling: generalizable re-identification in the operating room. Med. Image Anal., 103687 (2025)
30. Zhang, Y., et al.: ByteTrack: multi-object tracking by associating every detection box. In: European Conference on Computer Vision, pp. 1–21. Springer (2022)
31. Özsoy, E., et al.: MM-OR: a large multimodal operating room dataset for semantic understanding of high intensity surgical environments. In: CVPR (2025)

CryoFlow: Prediction of Frozen Region Growth in Kidney Cryoablation Using a 3D Flow-Matching

Siyeop Yoon[1], Yujin Oh[1], Matthew Tivnan[1], Sifan Song[1], Pengfei Jin[1], Sekeun Kim[1], Dufan Wu[1], Hyun Jin Cho[2], Raul Uppot[1], and Quanzheng Li[1]([✉])

[1] Department of Radiology and Center for Advanced Medical Computing and Analysis, Massachusetts General Hospital and Harvard Medical School, Boston, MA, USA
li.quanzheng@mgh.harvard.edu
[2] Chungnam National University Hospital, Chungnam National University College of Medicine, Daejeon, South Korea

Abstract. This study presents a 3D flow-matching model designed to predict the progression of the frozen region (iceball) during kidney cryoablation. Precise intraoperative guidance is critical in cryoablation to ensure complete tumor eradication while preserving adjacent healthy tissue. However, conventional methods, typically based on physics-driven or diffusion-based simulations, are computationally demanding and often struggle to accurately represent complex anatomical structures. To address these limitations, our approach leverages intraoperative CT imaging to inform the model. The proposed 3D flow-matching model is trained to learn a continuous deformation field that maps early-stage CT scans to future predictions. This transformation not only estimates the volumetric expansion of the iceball but also generates corresponding masks, effectively capturing spatial and morphological changes over time. Quantitative analysis highlights the model's robustness, showing strong agreement between predictions and ground-truth masks. The model achieves an Intersection over Union (IoU) score of 0.61 ± 0.11 and a Dice coefficient of 0.75 ± 0.11. By integrating real-time CT imaging with advanced deep learning techniques, this approach has the potential to enhance intraoperative guidance in kidney cryoablation, improving procedural outcomes and advancing the field of minimally invasive surgery. All codes are available https://github.com/siyeopyoon/CryoFlow.

Keywords: Data synthesis · Diffusion Models · Flow-Matching · Intervention · CT-guided · CT

1 Introduction

Kidney cryoablation is a minimally invasive treatment for small renal tumors that effectively destroys the tumor while preserving kidney function [13]. In this procedure, cryoprobes are inserted percutaneously under CT guidance to create an "iceball," a clearly defined area of frozen tissue that marks the ablation zone and protects nearby critical structures [6,19]. On CT images, the iceball appears as a low-attenuation (hypodense) region, usually with a spherical or ellipsoidal shape. Its well-defined borders are produced by ice crystal formation, which decreases the tissue density. Often, the medial edge of the iceball is shorter because the warmer blood flow in the central kidney limits the spread of freezing (the "cold sink" effect). Accurate, real-time prediction of the iceball's growth is critical to ensure that the entire tumor is covered with an adequate safety margin, thereby minimizing injury to healthy tissue [9].

Conventional strategies for predicting iceball dynamics have spanned from simple geometric approximations to sophisticated physics-based simulations, notably those employing the Pennes bioheat equation [4,14]. Although these methods can yield high accuracy, their substantial computational overhead and dependence on simplifying assumptions restrict their applicability in real-time intraoperative settings. Recent data-driven approaches, particularly those utilizing deep learning, have demonstrated promising improvements in prediction accuracy within analogous ablation contexts [10].

Moreover, advances in artificial intelligence have transformed medical image analysis; early generative frameworks, such as Variational Autoencoders and Generative Adversarial Networks, laid the foundation for realistic image synthesis [1,5], while modern diffusion-based models have significantly enhanced image fidelity and control over the generative process [2,3,12]. Even with these advances, significant challenges still exist—particularly the high computational costs and long processing times required for full 3D volumes. Techniques such as patch-wise processing [15,17] and working in latent space [11] have been suggested to ease these issues, but keeping the anatomical accuracy of the predictions remains difficult.

In our study, we utilize the principles of flow-matching models [7,8] to develop a novel 3D predictive framework for the future configuration of the ablation zone. Unlike conventional diffusion models that rely on iterative noise removal, our method directly learns a deformation field that continuously transforms early CT images into their predicted future states. This direct mapping preserves the high image quality and precise control characteristic of diffusion-based techniques, while significantly reducing computational requirements.

Furthermore, by incorporating patch-wise training strategies with the usage of a residual estimation approach, our model achieves an effective balance between computational efficiency and the retention of spatial and anatomical detail. Notably, as most anatomical structures remain unchanged during the procedure—with only the iceball exhibiting dynamic changes—our approach is specifically tailored to capture the evolving deformation of the ablation zone.

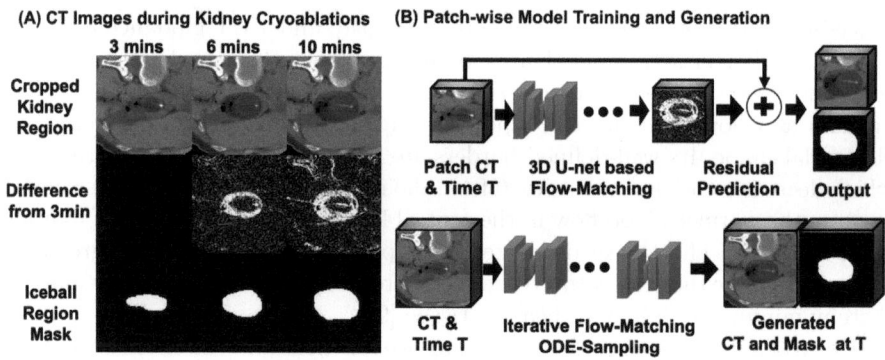

Fig. 1. (A) Data collection of 3D CT scans acquired at 3–10 min and their corresponding masks. (B) Patch-wise model training and generation framework, employing a 3D U-Net–based flow-matching approach with iterative ODE sampling to produce time-specific CT volumes and masks.

2 Methods

2.1 Data Description and Study Population

This study utilized a retrospective dataset of 24 patients who underwent CT-guided percutaneous renal cryoablation between September 2017 and May 2023. The study was approved by the Institutional Review Board (IRB), and informed consent was waived due to the retrospective nature of the study (Protocol: Anonymized). Each procedure involved 2–4 freeze–thaw cycles, with intraoperative CT images acquired at regular intervals (approximately every 2–3 min) during each freeze to monitor iceball formation. The complete dataset comprises 31 imaging studies, partitioned into training and testing sets on a per-patient basis to avoid data leakage. For testing, 8 cases were included, yielding 34 target scans acquired from the 3-min scan. Sixteen scans corresponded to the first cryoablation cycle, while 18 scans were obtained during the second cycle, with target scans acquired 6–10 min after the initial 3-min scan. All patients underwent CT-guided percutaneous renal cryoablation using devices manufactured by various vendors.

Model Inputs and Outputs. The proposed model accepts a structured tuple; specifically, the inputs consist of the source volume X, a stage indicator representing the current freeze cycle phase, and time change information (e.g., elapsed time or Δt). The trained model simultaneously predicts both the future CT image and the corresponding iceball mask, as shown in the Fig. 1. The output is a tuple comprising the predicted target CT volume Y and a binary mask delineating the iceball. Joint optimization for image synthesis and mask ensures that the network effectively captures both the appearance and spatial extent of the evolving iceball.

Flow-Matching for Residual Transformation. To predict the future state of the frozen region, we employ a 3D flow-matching model that learns a continuous mapping from an early CT image, denoted by I_{src} (e.g., acquired at 3 min), to a predicted future CT image, denoted by I_{tgt} (e.g., acquired at 10 min), while simultaneously predicting the corresponding iceball mask at the future time point.

Let $I_{\text{src}} \in \mathbb{R}^{H \times W \times D}$ denote the source CT volume and $I_{\text{tgt}} \in \mathbb{R}^{H \times W \times D}$ denote the paired target volume. We define the residual volume as

$$r = I_{\text{tgt}} - I_{\text{src}}.$$

To model the transformation from I_{src} to I_{tgt}, we introduce a normalized interpolation parameter $\tau \in [0, 1]$, such that the intermediate volume is given by

$$I(\tau) = I_{\text{src}} + \tau r = I_{\text{src}} + \tau (I_{\text{tgt}} - I_{\text{src}}).$$

Thus, $I(0) = I_{\text{src}}$ and $I(1) = I_{\text{tgt}}$. Because the interpolation is linear, the derivative with respect to τ is constant, $\frac{dI(\tau)}{d\tau} = r$. This implies that the true residual velocity field is $u(\tau) = r, \quad \forall \tau \in [0, 1]$. A neural network $\mathcal{N}_\theta$ is employed to estimate the residual velocity field from the intermediate image $I(\tau)$ and the interpolation parameter τ, yielding the prediction

$$\hat{u}(\tau) = \mathcal{N}_\theta\big(I(\tau), \tau\big). \tag{1}$$

The training objective minimizes the mean squared error (MSE) between the predicted velocity field and the true residual:

$$\mathcal{L}(\theta) = \mathbb{E}_{(I_{\text{src}}, I_{\text{tgt}}), \tau \sim \mathcal{U}(0,1)} \left[\left\| \mathcal{N}_\theta\big(I_{\text{src}} + \tau (I_{\text{tgt}} - I_{\text{src}}), \tau\big) - (I_{\text{tgt}} - I_{\text{src}}) \right\|^2 \right]. \tag{2}$$

To further increase data diversity and computational efficiency, the model is also trained in a patch-wise manner. Three-dimensional patches of size $32 \times 32 \times 32$ are extracted from paired CT volumes and their corresponding iceball masks. Let P_{src} and P_{tgt} denote a pair of patches extracted from two temporally paired volumes. The residual patch is defined as

$$r_{\text{patch}} = P_{\text{tgt}} - P_{\text{src}}, \text{ and } P(\tau) = P_{\text{src}} + \tau r_{\text{patch}}, \quad \tau \in [0, 1], \tag{3}$$

so that $P(0) = P_{\text{src}}$ and $P(1) = P_{\text{tgt}}$. Under this formulation, the derivative is constant:

$$\frac{dP(\tau)}{d\tau} = r_{\text{patch}}.$$

The patch-wise flow-matching loss is then given by

$$\mathcal{L}_{\text{patch}}(\theta) = \frac{1}{N} \sum_{i=1}^{N} \left\| \mathcal{N}_\theta\big(P(\tau_i), \tau_i\big) - r_{\text{patch},i} \right\|^2, \tag{4}$$

where N is the number of patch pairs in the batch and each τ_i is sampled uniformly from $[0, 1]$. The trained 3D U-Net is used to predict the future CT image by integrating the estimated velocity field over τ:

$$I_{\text{tgt}} = I_{\text{src}} + \int_0^1 \mathbf{v}(x, \tau) \, d\tau. \tag{5}$$

This integration is performed using an ODE solver (e.g., the Heun 2nd order solver). In our implementation, the mask is provided as an additional input channel to the network. This enables simultaneous prediction of the future iceball mask; thus, the overall training objective is the patch-wise flow-matching loss calculated from 2-channel volume data.

Finally, the 3D flow-matching model produces two outputs: a synthetic CT volume representing the kidney and the evolving iceball at a future time point, and a mask of the frozen region. We threshold mask channel with 0.95 to generate binary masks. Additionally, the framework allows estimation of the continuous growth of the iceball by controlling a stage indicator representing the current freeze cycle phase and time change information, which are provided as an additional input channel to the model.

2.2 Data Preprocessing and Model Implementation

For the training, CT images were preprocessed to ensure consistent spatial representation and robust masking. First, all volumes were resampled to an isotropic resolution of $1\,\text{mm}^3$. Next, within each freeze cycle, images were rigidly registered to compensate for respiratory motion using ITK [16]. The frozen region was segmented using ITK-SNAP [18]: a semi-automatic approach yielded initial ground truth labels, which were further refined via manual segmentation to include the catheter. For optimal delineation, the CT visualization window was set to $[-150, 350]$ HU. Finally, random 90,180,270° rotations were applied during training to improve model generalization by considering various catheter insertion locations and directions during the procedure.

The training dataset was constructed by randomly pairing CT scans acquired at different time points within the same cryoablation cycle. For instance, scans acquired at 6 min were paired with those at 10 min (time difference = +4 min), and scans at 3 min (time difference = −3 min). This strategy augments the number of training pairs and introduces pairs with negative time differences, artificially reflecting a reverse cryoablation process.

The magnitude-preserving 3D U-Net was used as a residual velocity field estimator by minimizing patch-wise loss, Eq. 4. We extended the convolutional layers of the EDM2 model to 3D and deactivated the preconditioning step used in the original implementation. The resulting model effectively preserves the magnitude of the feature representations throughout the network. Training was performed with the following hyperparameter settings (duration = 12k gradient update iterations, batch = 2048, channels = 64, learning rate = 0.0170, learning rate half-life = 35000 on 8 A100 GPUs over a 24-h period). For comparative

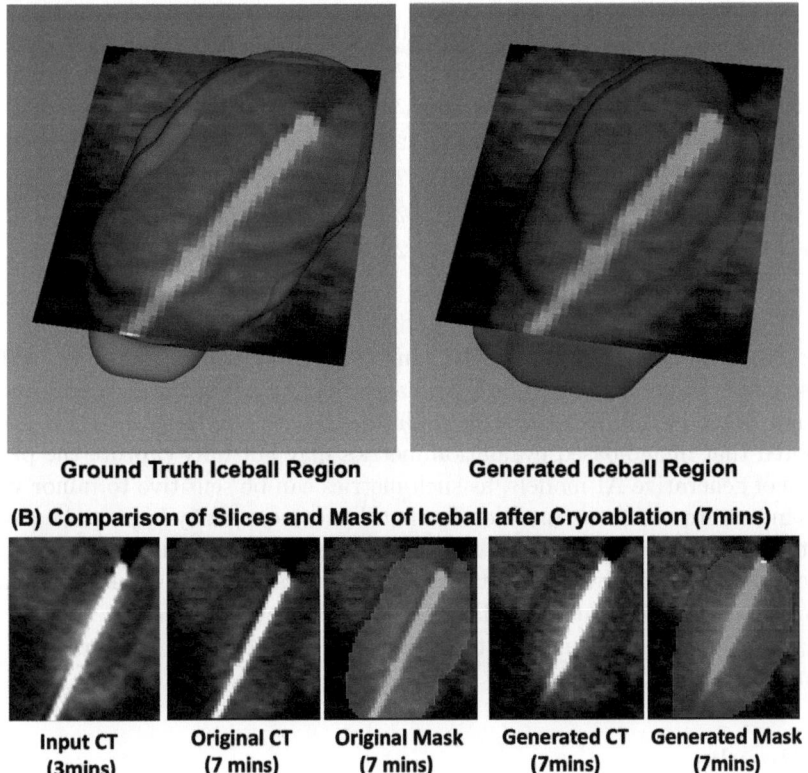

Fig. 2. (A) Comparison of 3D surface models of the Iceball region, contrasting the ground truth (green) with the generated region (red) overlaid on axial CT slices. (B) Slice-by-slice comparison of the Iceball region masks, illustrating input and generated CTs alongside corresponding ground-truth and generated masks. (Color figure online)

analysis, a diffusion model was also evaluated using the same settings with an EDM sampler employing 10–100 steps. The full model comprises approximately 80 million parameters and employs mixed-precision (FP16) training to accelerate computation. Under these conditions, both the predicted CT volume and the iceball mask are generated in approximately **15 s per inference** on our hardware (single A100 GPU) with a cropped volume around the cryoablation catheter (size of 64^3). All codes are available https://github.com/siyeopyoon/CryoFlow.

2.3 Evaluation

Our evaluation comprises both quantitative metrics and qualitative visualizations to assess the performance of the proposed flow-matching model for predicting iceball evolution. For quantitative evaluation, image quality was measured using standard metrics including normalized Mean Absolute Error (NMAE),

Table 1. Performance evaluation of Iceball Region and Mask Prediction.

Method	NMAE (x10)	PSNR (dB)	SSIM	IoU	Dice
Diffusion (10)	0.87 ± 0.2	18.67 ± 2.19	0.360 ± 0.14	0.59 ± 0.11	0.73 ± 0.09
Diffusion (50)	0.82 ± 0.2	19.17 ± 2.40	0.400 ± 0.15	0.56 ± 0.14	0.70 ± 0.13
Diffusion (100)	0.83 ± 0.2	19.13 ± 2.39	0.403 ± 0.15	0.55 ± 0.11	0.71 ± 0.09
CryoFlow (10)	0.82 ± 0.2	19.23 ± 2.35	0.403 ± 0.15	0.57 ± 0.13	0.72 ± 0.11
CryoFlow (50)	0.82 ± 0.2	19.21 ± 2.48	0.404 ± 0.15	0.59 ± 0.11	0.73 ± 0.10
CryoFlow (100)	0.80 ± 0.2	19.32 ± 2.32	0.410 ± 0.15	0.61 ± 0.11	0.75 ± 0.11

Peak Signal-to-Noise Ratio (PSNR), and Structural Similarity Index (SSIM). These metrics were computed on CT images that were first normalized from the original range $[-150, 350]$ Hounsfield Units (HU) to the range $[0, 1]$. It should be noted that image-based evaluation metrics may not fully capture the performance of generative AI models, as such metrics can be sensitive to minor variations in texture and structural details that do not necessarily impact the clinical utility of the synthesized images. Therefore, quantitative metrics should be interpreted in conjunction with qualitative visual assessments. For mask generation performance, we calculated the Dice Similarity Coefficient (DSC), Intersection over Union (IoU), and the absolute volume difference between the predicted and ground truth masks.

3 Results

The representative 3D surface comparison reveals that the generated Iceball region closely follows the contour and volume of the ground truth, showing minimal discrepancies in shape and boundary (Fig. 2). The slice-wise masks further confirm this accuracy, as the generated mask exhibits strong agreement with the original mask across multiple cross-sections. These findings indicate that the proposed method effectively captures both global and local features of the ablation zone, resulting in precise reconstructions of the Iceball region.

The quantitative evaluation of the Iceball region prediction and mask is summarized in Table 1. Two methods—Diffusion and Flow models-were assessed across three sampling step settings (10, 50, and 100 iterations). For the Diffusion method, the performance at 10 iterations shows an NMAE of 0.87, a PSNR of 18.67dB, an SSIM of 0.360, an IoU of 0.59, and a Dice coefficient of 0.73. Increasing the iterations to 50 and 100 yields modest improvements in PSNR and SSIM, with values around 19.17dB/0.400 and 19.13dB/0.403 respectively. However, the IoU and Dice scores do not exhibit a consistent enhancement, indicating that higher iterations may refine certain image quality aspects without a proportional gain in mask generation accuracy.

In contrast, the Flow method demonstrates a clear trend of improvement with increased iterations. At 10 iterations, Flow achieves an NMAE of 0.82, a PSNR of

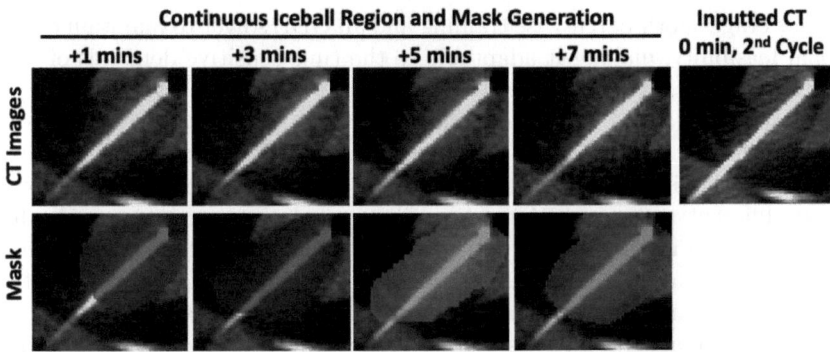

Fig. 3. Continuous iceball region and mask generation for a second cryoablation cycle, starting from an input CT at 0 min. Each column depicts the predicted CT slice (top row) and its corresponding iceball mask (bottom row) at incremental time points of +1, +3, +5, and +7 min.

19.23dB, an SSIM of 0.403, an IoU of 0.57, and a Dice of 0.72. This performance further improves at 50 iterations and reaches its peak at 100 iterations—with an NMAE of 0.80, a PSNR of 19.32dB, an SSIM of 0.410, an IoU of 0.61, and a Dice coefficient of 0.75. These results indicate that the Flow-based approach is particularly effective in enhancing both region prediction accuracy and mask quality when provided with a higher number of iterations.

For qualitative evaluation, we synthesized CT images at two-min intervals starting from the starting image at the second cryoablation cycle. These continuous mapping visualizations (see Fig. 3) clearly illustrate the progressive evolution of the iceball. Each synthesized image shows incremental changes in the frozen region, capturing the subtle deformations and growth patterns over time. The smooth, gradual changes observed in the synthesized images and mask predictions highlights the model's capacity to effectively capture and predict the continuous evolution of the iceball.

4 Discussion

This work introduces a 3D flow-matching model for predicting the progression of frozen regions in kidney cryoablation. By leveraging patch-wise training, the approach efficiently handles volumetric data while retaining spatial details. Comparative evaluations show that the flow-matching framework consistently outperforms a diffusion-based baseline in both image-quality metrics and mask generation accuracy, especially when employing higher iteration steps.

The proposed 3D flow-matching model offers a data-driven approach to predicting the spatial evolution of frozen regions in kidney cryoablation. By learning a continuous deformation field that connects early and later time points, the method directly models local changes in the ablation zone without requiring complex physics-based simulations or iterative noise-removal steps. This design choice leverages both patch-wise training and an ODE solver to maintain computational feasibility, making it adaptable to the time-sensitive demands of intraoperative settings.

Nevertheless, a few limitations should be noted. First, the retrospective dataset, although diverse in terms of imaging intervals and tumor sizes, is drawn from a single institution. As a result, generalization to other clinical contexts—such as different scanner settings or patient populations—remains to be fully explored. Future research could extend this flow-matching framework by incorporating additional physiological variables (e.g., blood perfusion or tissue thermal properties) and exploring multi-task learning setups that predict not only the iceball boundary but also associated tissue viability or necrosis. Real-time or near-real-time inference can be further facilitated by optimizing network architectures and parallelizing the ODE-solver steps. Finally, prospective validation on larger, multicenter datasets is warranted to confirm the clinical utility and robustness of this approach.

Disclosure of Interests. The authors have no competing interests to declare that are relevant to the content of this article.

References

1. Goodfellow, I., et al.: Generative adversarial networks. Commun. ACM **63**(11), 139–144 (2020)
2. Ho, J., Jain, A., Abbeel, P.: Denoising diffusion probabilistic models. In: Advances in Neural Information Processing Systems, vol. 33, pp. 6840–6851 (2020)
3. Karras, T., Aittala, M., Aila, T., Laine, S.: Elucidating the design space of diffusion-based generative models. In: Advances in Neural Information Processing Systems, vol. 35, pp. 26565–26577 (2022)
4. Kim, C., O'Rourke, A.P., Mahvi, D.M., Webster, J.G.: Finite-element analysis of Ex vivo and in vivo hepatic cryoablation. IEEE Trans. Biomed. Eng. **54**(7), 1177–1185 (2007)
5. Kingma, D.P., Welling, M., et al.: Auto-encoding variational Bayes (2013)
6. Knox, J., et al.: Intermediate to long-term clinical outcomes of percutaneous cryoablation for renal masses. J. Vasc. Interv. Radiol. **31**(8), 1242–1248 (2020)
7. Lipman, Y., Chen, R.T., Ben-Hamu, H., Nickel, M., Le, M.: Flow matching for generative modeling. arXiv preprint arXiv:2210.02747 (2022)
8. Lipman, Y., et al.: Flow matching guide and code. arXiv preprint arXiv:2412.06264 (2024)
9. Littrup, P.J., et al.: Lethal isotherms of cryoablation in a phantom study: effects of heat load, probe size, and number. J. Vasc. Interv. Radiol. **20**(10), 1343–1351 (2009)
10. Moreira, P., Tuncali, K., Tempany, C., Tokuda, J.: AI-based isotherm prediction for focal cryoablation of prostate cancer. Acad. Radiol. **30**, S14–S20 (2023)
11. Rombach, R., Blattmann, A., Lorenz, D., Esser, P., Ommer, B.: High-resolution image synthesis with latent diffusion models. In: Proceedings of the IEEE/CVF Conference on Computer Vision and Pattern Recognition, pp. 10684–10695 (2022)

12. Song, Y., Sohl-Dickstein, J., Kingma, D.P., Kumar, A., Ermon, S., Poole, B.: Score-based generative modeling through stochastic differential equations. arXiv preprint arXiv:2011.13456 (2020)
13. Stacul, F., et al.: Cryoablation of renal tumors: long-term follow-up from a multicenter experience. Abdom. Radiol. **46**(9), 4476–4488 (2021). https://doi.org/10.1007/s00261-021-03082-z
14. Tanwar, S., Famhawite, L., Verma, P.R.: Numerical simulation of bio-heat transfer for cryoablation of regularly shaped tumours in liver tissue using multiprobes. J. Therm. Biol **113**, 103531 (2023)
15. Wang, Z., et al.: Patch diffusion: faster and more data-efficient training of diffusion models. arXiv preprint arXiv:2304.12526 (2023)
16. Yoo, T.S., et al.: Engineering and algorithm design for an image processing API: a technical report on ITK-the insight toolkit. In: Medicine Meets Virtual Reality 02/10, pp. 586–592. IOS Press (2002)
17. Yoon, S., et al.: High-resolution 3D CT synthesis from bidirectional X-ray images using 3D diffusion model. In: 2024 IEEE International Symposium on Biomedical Imaging (ISBI), pp. 1–4. IEEE (2024)
18. Yushkevich, P.A., Gao, Y., Gerig, G.: ITK-SNAP: an interactive tool for semi-automatic segmentation of multi-modality biomedical images. In: 2016 38th Annual International Conference of the IEEE Engineering in Medicine and Biology Society (EMBC), pp. 3342–3345. IEEE (2016)
19. Zhou, W., Herwald, S.E., McCarthy, C., Uppot, R.N., Arellano, R.S.: Radiofrequency ablation, cryoablation, and microwave ablation for T1A renal cell carcinoma: a comparative evaluation of therapeutic and renal function outcomes. J. Vasc. Interv. Radiol. **30**(7), 1035–1042 (2019)

Temporally Stable Monocular Depth Estimation in Surgical Vision

Jialang Xu[1,2,3](✉), Emanuele Colleoni[1], Nicolas Toussaint[1], Muhammad Asad[1], Ricardo Sanchez-Matilla[1], Evangelos B. Mazomenos[2,3], Imanol Luengo[1], and Danail Stoyanov[1,2,4]

[1] Medtronic, London, UK
[2] UCL Hawkes Institute, University College London, London, UK
jialang.xu.22@ucl.ac.uk
[3] Department of Medical Physics and Biomedical Engineering, University College London, London, UK
[4] Department of Computer Science, University College London, London, UK

Abstract. Recent foundational models have unlocked numerous possibilities for computer-assisted interventions. A critical advancement in this field is the precise estimation of dense relative depth on surgical videos, essential for understanding the 3D positioning of surgical instruments and measuring anatomical structures. However, existing methods often struggle to estimate depth maps that are coherent and smooth over time, leading to noisy and temporally inconsistent depth predictions. We propose TAN, a novel Temporal Adapter Network for monocular depth estimation that enhances the foundational model Depth Anything V2 from image-based to temporally aware depth estimation. Specifically, we design a lightweight temporal adapter and integrate it into the decoder to capture temporal features from consecutive frames. Additionally, we introduce a self-supervised temporal regularization loss, utilizing optical flow to enforce stable depth estimation between consecutive frames. Our experiments, conducted on the SCARED and EndoNeRF datasets, two established benchmarks for evaluating depth estimation models in the surgical domain, demonstrate that the proposed TAN improves both temporal consistency and depth accuracy, achieving at least a 14.29% reduction in OPW and 3.6% in RMSE on SCARED, and 6.2% in OPW and 3.26% in RMSE on EndoNeRF compared to state-of-the-art methods, while running at 97 FPS, making it well-suited for real-time surgical applications.

Keywords: Monocular Depth Estimation · Temporal Consistency · Surgery

1 Introduction

Monocular depth estimation (MDE) plays a crucial role in surgical applications, including intra-operative navigation, integration of augmented reality, and scene reconstruction [6,15]. However, traditional multi-view methods such as structure-from-motion (SfM) [14] and simultaneous localization and mapping [5] struggle to work in surgical environments due to low lighting conditions and textureless surfaces [8,20]. In this context, deep learning has been applied to advance surgical MDE. For instance, Faisal et al. [17] proposed a deep convolutional network with a conditional random field framework for endoscopy depth estimation, while AF-SfMLearner [20] introduced an appearance flow constraint to address brightness inconsistencies across frames, reducing severe inter-frame fluctuations and enhancing depth estimation accuracy.

Recently, foundation models, trained on vast datasets with millions or even billions of samples, have achieved state-of-the-art performance in various tasks [4,8,18]. Notable examples include the general-purpose visual feature extraction model DINOv2 [18], the segmentation model Segment Anything [13], and depth estimation models such as Depth Anything (DAv1) [25] and Depth Anything V2 (DAv2) [26]. Inspired by the remarkable performance of foundation models in downstream tasks, more recent efforts have focused on their adaptation for surgical MDE. Surgical-DINO [7] employs low-rank adaptation (LoRA) [11] to fine-tune the frozen DINOv2 [18] encoder with a custom-designed depth decoder, while EndoDAC [8] integrates dynamic vector-based LoRA and self-supervised camera intrinsics estimation to adapt DAv1 [25] for surgical MDE. Though these methods excel in per-frame depth accuracy, they are mainly designed for single-image depth estimation and suffer from inconsistent depth estimates over time. Temporal consistency refers to the stability and coherence of depth predictions across consecutive video frames. In a temporally consistent model, depth variations arise primarily from genuine scene or camera motion, rather than random fluctuations or noise introduced by the model. In surgical environments, temporally inconsistent depth predictions can be visually distracting and cognitively taxing for the operating surgeon. Such instability may disrupt real-time navigation systems or augmented reality overlays, potentially increasing the risk of misinterpretation and compromising patient safety. Ensuring smooth and reliable depth estimation is therefore critical for maintaining surgical workflow and supporting intra-operative decision-making.

We propose a temporal adapter network (TAN) based on DAv2 for MDE, designed to enhance both depth accuracy and temporal consistency in surgical depth estimation. Our key contributions are as follows:

- We propose TAN, a novel method that efficiently adapts prior knowledge from DAv2 [26] and enables temporal information interaction across frames. By innovating a series of lightweight temporal adapters, TAN requires only 3.1 million trainable parameters to achieve temporally stable and accurate depth predictions in surgical videos.
- To enforce temporal consistency, we introduce a self-supervised temporal regularization term into the depth loss, ensuring alignment of depth predictions across consecutive frames during the training process.
- We provide extensive experiments on two publicly available depth datasets to demonstrate that TAN outperforms state-of-the-art MDE approaches in both

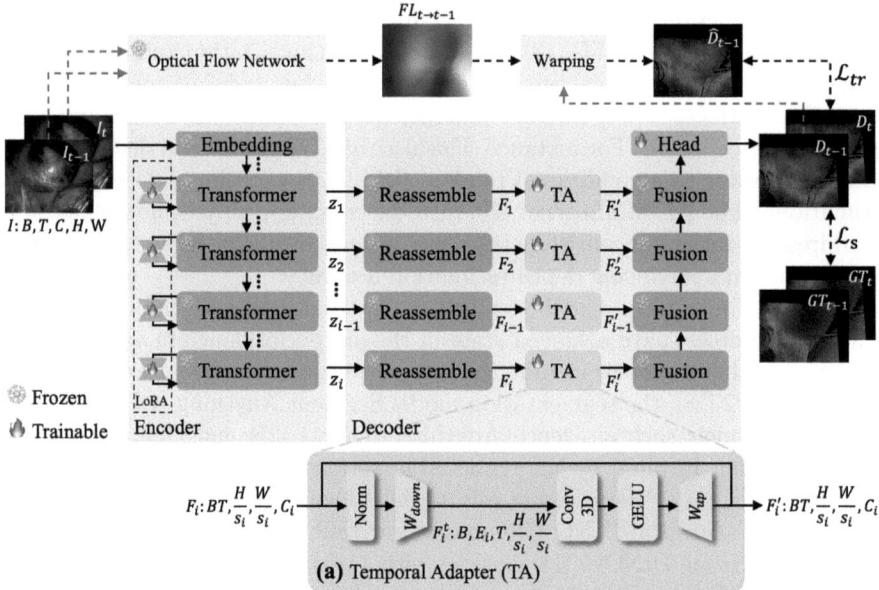

Fig. 1. Architecture of the proposed Temporal Adapter Network (TAN), showing LoRA-enhanced encoder, temporal adapters in the decoder, and training-time optical flow regularization. The image encoder is updated through LoRA. Temporal information is incorporated via a set of temporal adapters (TAs). The depth head of the decoder is also trainable. Dashed arrows indicate paths used during training but omitted during inference. (a) Proposed TA.

depth accuracy and temporal stability, while maintaining high computational efficiency for real-time applications.

2 Methodology

2.1 Depth Anything

DAv1 [25] and DAv2 [26] are models developed to perform MDE in general contexts. They share the same architecture, which includes a Vision Transformer (ViT)-based image encoder [1] and a dense prediction Transformer (DPT) decoder [19]. The DPT decoder utilizes four Reassemble layers to construct a feature pyramid, four fusion layers to progressively refine feature maps from low to high resolution, and a depth head with two convolutional layers to output depth predictions.

2.2 Temporal Adapter Network

The proposed TAN enhances the DAv2 foundation model by incorporating temporal consistency, as shown in Fig. 1. Building on DAv2's architecture, TAN comprises three main components: (i) low-rank adaptation (LoRA) [11] enhancing the ViT-based encoder, (ii) a novel temporal adapter (TA) embedded within the DPT decoder, (iii) a self-supervised temporal regulation loss $\mathcal{L}_{tr}$ optimized alongside a depth loss $\mathcal{L}_s$.

Given a sequence of consecutive frames $I \in \mathbb{R}^{B \times T \times C \times H \times W}$, we collapse the temporal dimension into the batch dimension, yielding $I \in \mathbb{R}^{(B \times T) \times C \times H \times W}$, where B, T, C, H, and W denote the batch size, number of frames, number of channels, height, and width, respectively. The image encoder extracts spatial tokens $z_i \in \mathbb{R}^{(B \times T) \times N \times D_i}$ from four different Transformer blocks, where $i \in \{1, 2, 3, 4\}$, $N = \frac{H}{p} \times \frac{W}{p}$ is the number of spatial patches, p is the patch size, and D_i is the embedding dimension. The Reassemble layers [19] in the decoder restructure and resample z_i into spatial feature maps $F_i \in \mathbb{R}^{(B \times T) \times \frac{H}{s_i} \times \frac{W}{s_i} \times C_i}$, where s_i is the resampling scale and C_i is the number of channels. Following [11], since DAv2 is pretrained on real and synthetic natural images, we integrate LoRA into the query and value projections of the multi-headed self-attention mechanism, enabling adaptation to surgical images and improving spatial representations for subsequent temporal modeling. Notably, while the image encoder and Reassemble layers can capture spatial representations for a single frame, they do not account for temporal dependencies, leading to inconsistencies in depth predictions over time.

To enforce temporal consistency, we introduce two complementary mechanisms: (i) four temporal adapters inserted between the Reassemble and Fusion layers of the DPT-based decoder to explicitly model inter-frame dependencies; and (ii) augmenting depth supervision with $\mathcal{L}_{tr}$, which smooths depth values across consecutive frames. During training, only the LoRA, temporal adapters, and the depth head of the decoder are updated, while all other parameters remain frozen. During inference, the optical flow network is discarded. Details of the two innovations follow.

Temporal Adapter. Once spatial features are extracted, TAN focuses on capturing inter-frame correlations to model temporal dependencies via the proposed TA. As illustrated in Fig. 1(a), the TA consists of: a normalization layer, a down-projection linear layer, a 3D convolutional layer, a GELU activation [10], and an up-projection linear layer. Given a spatial feature map $F_i \in \mathbb{R}^{(B \times T) \times \frac{H}{s_i} \times \frac{W}{s_i} \times C_i}$ from the Reassemble layer, we first apply down-projection along the feature dimension C_i to E_i, yielding a compact representation that reduces computational cost. The temporal dimension is then separated as $F_i^t \in \mathbb{R}^{B \times E_i \times T \times \frac{H}{s_i} \times \frac{W}{s_i}}$, and a 3D convolution is applied to extract temporal information across T-dimension. Finally, the up-projection restores the feature dimension from E_i to C_i. A residual connection ensures the preservation of original spatial information. Formally, TA is defined as:

$$\text{TA}(F_i) = F_i + W_{up}\delta\left(\text{Conv3D}\left(W_{down}\left(\text{Norm}(F_i)\right)\right)\right) \quad (1)$$

where Norm is the layer normalization, $W_{down} \in \mathbb{R}^{C_i \times E_i}$ and $W_{up} \in \mathbb{R}^{E_i \times C_i}$ are down- and up-projection layers, respectively. Conv3D is the 3D convolutional layer with kernel size (3, 1, 1), and δ is the GELU activation function.

Self-supervised Temporal Regulation Loss. To further improve temporal consistency, we introduce a self-supervised temporal regulation loss $\mathcal{L}_{tr}$. As shown in Fig. 1, given two adjacent frames I_{t-1} and I_t, we obtain depth predictions D_{t-1} and D_t via TAN. An optical flow network establishes correspondences between I_{t-1} and I_t, aligning D_{t-1} to time t. In this paper, we adopt GMFlow [24] as the optical flow network, leveraging its robustness, high performance and computational efficiency. The temporal regulation loss $\mathcal{L}_{tr}$ promotes smooth depth transitions across frames and is defined as:

$$\mathcal{L}_{tr}(t, t-1) = \frac{1}{n} \sum_{i=1}^{n} M_{t \to t-1}^{(i)} \|D_t^{(i)} - \hat{D}_{t-1}^{(i)}\|_1 \quad (2)$$

where n is the number of pixels per frame, $\|\cdot\|_1$ denotes the L_1 distance, $\hat{D}_{t-1}$ is the predicted depth D_{t-1} warped by the optical flow $FL_{t \to t-1}$, $FL_{t \to t-1}$ is the backward flow from I_{t-1} to I_t; and $M_{t \to t-1} = e^{(-50\|I_t - \hat{I}_{(t-1)}\|_2^2)}$ is the visibility mask accounting for occlusions [3,21,22].

The complete training loss for TAN is

$$\mathcal{L}_{total} = \mathcal{L}_s + \beta \mathcal{L}_{tr} \quad (3)$$

where $\mathcal{L}_s$ is the scale-invariant log loss [9] for depth supervision, and β controls the weight of the temporal consistency term.

3 Experiments

SCARED Dataset. [2] consists of nine video recordings of fresh porcine cadaver abdominal anatomy, each acquired at 25 frames per second (FPS) with a da Vinci Xi surgical robot. Ground-truth depth maps are obtained with a structured light projector. Previous works [7,8,20] used an unordered test split, which does not allow for proper temporal consistency evaluation. To address this, we follow the original dataset split scheme proposed in [2], using 22,958 frames (videos 1–7) for training and 5,909 frames (videos 8–9) for testing.

EndoNeRF Dataset. [23] includes two video clips (219 frames in total) from da Vinci robotic prostatectomy procedures. Ground-truth depth maps are estimated using a stereo depth estimation model [16]. Videos are divided in 4–8 s clips, recorded at 15 FPS. The first clip, consisting of 63 frames, shows tissue pushing/pulling and is used for training, while the second clip, consisting of 156 frames, depicts tissue cutting and is used for testing.

Evaluation Metrics. We evaluate both temporal consistency and per-frame depth accuracy. To measure temporal consistency, we employ the optical flow-based warping (OPW) and relative temporal consistency (RTC) metrics, following [12,21,22]. OPW is the average change in the depth of all pixel points

Table 1. Comparison of depth estimation methods on SCARED and EndoNeRF datasets. DAv2-ZS and DAv2-FT denote DAv2 in zero-shot and fine-tuned settings, respectively. OPW: Optical flow-based warping error; RTC: Relative temporal consistency; δ_1: Threshold accuracy. Top two results are highlighted in **bold** and <u>underline</u>.

	Method	OPW ↓	RTC ↑	Abs Rel ↓	Sq Rel ↓	RMSE ↓	RMSE$_{log}$ ↓	δ_1 ↑
SCARED	AF-SfMLearner [20]	<u>0.3799</u>	<u>0.7379</u>	0.0758	0.3911	3.3477	0.0995	0.9589
	EndoDAC [8]	0.6273	0.5117	0.0884	0.4818	3.8883	0.1144	0.9452
	EndoDAC$_{DAv2}$ [8]	0.4951	0.6356	0.0861	0.4829	3.8371	0.1131	0.9424
	DAv2-ZS [26]	0.8665	0.3425	0.0841	0.4914	3.8365	0.1128	0.9222
	DAv2-FT [26]	0.7561	0.3537	0.0662	0.2888	3.0205	0.0889	0.9656
	Surgical-DINO [7]	0.6347	0.4511	<u>0.0590</u>	<u>0.2623</u>	<u>2.7570</u>	<u>0.0802</u>	<u>0.9684</u>
	TAN (ours)	**0.3256**	**0.7426**	**0.0577**	**0.2346**	**2.6577**	**0.0787**	**0.9739**
EndoNeRF	AF-SfMLearner [20]	3.4108	0.4785	0.1707	2.5148	11.2502	0.2075	0.7717
	EndoDAC [8]	4.8717	0.4811	0.2287	5.2231	17.0360	0.2619	0.6248
	EndoDAC$_{DAv2}$ [8]	<u>2.8181</u>	<u>0.6199</u>	0.1767	3.0007	12.5270	0.2259	0.7507
	DAv2-ZS [26]	3.8931	0.4410	0.2106	3.5048	14.7483	0.2470	0.5350
	DAv2-FT [26]	3.2604	0.5204	0.1797	2.7196	11.1974	0.2208	0.7441
	Surgical-DINO [7]	3.2553	0.5110	<u>0.1621</u>	**2.1828**	<u>10.4142</u>	<u>0.1935</u>	<u>0.7939</u>
	TAN (ours)	**2.6412**	**0.6395**	**0.1555**	<u>2.2448</u>	**10.0746**	**0.1904**	**0.8062**

across frames, while RTC evaluates the relative depth change between adjacent frames to ensure smoothness consistent with human perceptual judgments. Following [8,20], for single-frame depth accuracy, we adopt five standard metrics: absolute relative error (Abs Rel), squared relative error (Sq Rel), root mean squared error (RMSE), log RMSE (RMSE$_{log}$), and threshold accuracy (δ_1).

Implementation Details. All experiments are conducted using PyTorch on an NVIDIA RTX A6000 GPU. We use AdamW optimizer with an initial learning rate of $5 \cdot 10^{-5}$, a batch size of 8, a total of 50 epochs, and $T = 2$ frames per sequence. Following DAv2, we use a polynomial decay learning rate schedule with a decay exponent of 0.9. We initialize TAN with the official DAv2-Base weights [26]. We set $\beta = 0.5$ and $E_i = 384$. During evaluation, we apply median scaling [7,8,20] to rescale the predicted depth maps.

Comparison with the State-of-the-Art. We compare our method with several state-of-the-art monocular depth estimation methods designed for surgical vision [7,8,20] and natural vision [26]. These models are implemented from released code and original literature, and fine-tuned on the SCARED and EndoNeRF datasets. Note that, as the training and testing splits of EndoDAC [8] differ from ours, the pretrained Depth Anything model provided by EndoDAC has likely used data from our test set during training. To address this concern, we initialize the EndoDAC network using the official DAv1-Base weights [25] before fine-tuning. Additionally, we also utilize official DAv2-Base weight [26]

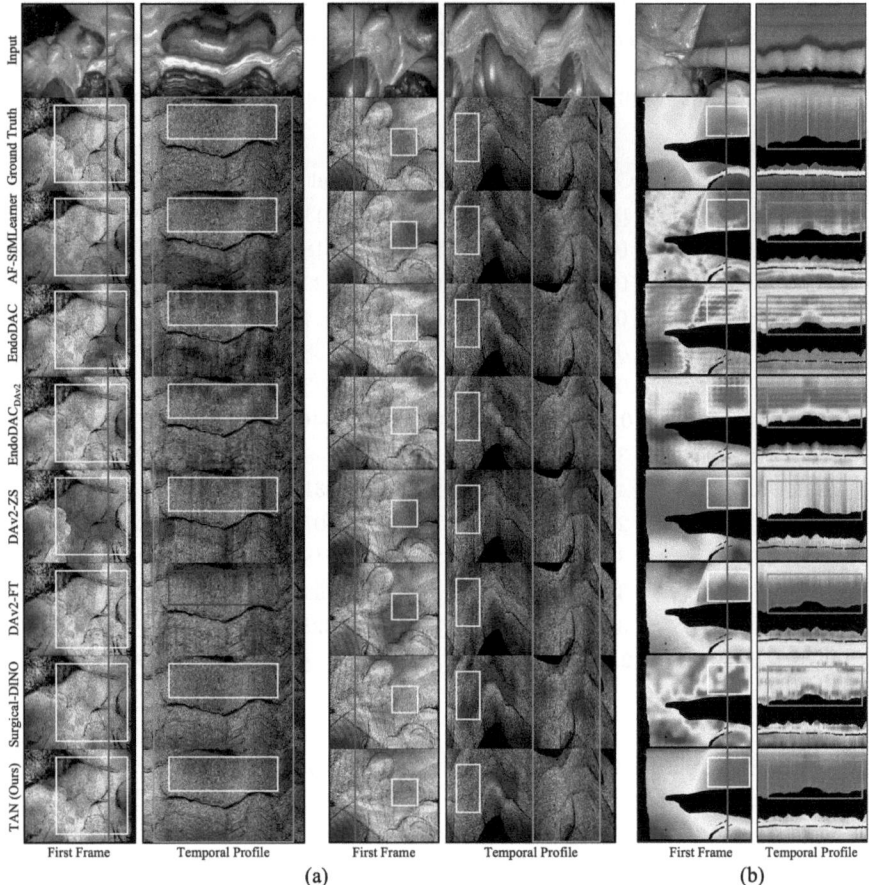

Fig. 2. Qualitative comparison of depth estimation methods on two test datasets: (a) SCARED, (b) EndoNeRF. The first column shows the first frame of the test sequence, while the second column (Temporal Profile) shows the changes in both input and depth over time at the vertical red line. White boxes show regions where our method has higher accuracy, while blue boxes highlight superior temporal consistency. Temporal profiles visualize depth stability across frames; fewer misaligned stripes indicate better temporal consistency. (Color figure online)

for initializing EndoDAC, denoted as EndoDAC$_{DAv2}$, to ensure a fair comparison. As shown in Table 1, TAN consistently outperforms other methods across all metrics on SCARED dataset and most of the metrics on EndoNeRF dataset, demonstrating superior temporal consistency and depth accuracy. Specifically, compared to state-of-the-art methods, our TAN reduces OPW by 14.29% and Sq Rel by 10.56% on the SCARED dataset, and achieves 6.28% lower OPW and 4.07% lower Abs Rel on the EndoNeRF dataset.

Table 2. Ablation study on SCARED dataset showing the impact of Temporal Adapter (TA) and Temporal Regulation Loss ($\mathcal{L}_{tr}$). OPW: Optical flow-based warping error; RTC: Relative temporal consistency; δ_1: Threshold accuracy; #TP is the number of trainable parameters. M = 10^6. Top result highlighted in **bold**.

TA	$\mathcal{L}_{tr}$	OPW ↓	RTC ↑	Abs Rel ↓	Sq Rel ↓	RMSE ↓	RMSE$_{log}$ ↓	δ_1 ↑	#TP (M)
Baseline		0.6541	0.4040	0.0670	0.3247	3.0382	0.0903	0.9592	0.3
✓		0.4967	0.4477	0.0587	0.2580	2.7045	0.0798	0.9728	3.1
	✓	0.3327	0.7234	0.0619	0.2729	2.9436	0.0856	0.9681	0.3
✓	✓	**0.3256**	**0.7426**	**0.0577**	**0.2346**	**2.6577**	**0.0787**	**0.9739**	3.1

Figure 2 presents depth estimation visualizations across long consecutive frames on two datasets, including the first frame of each video clip and its corresponding temporal profile. The temporal profile is generated by slicing frames along the timeline at the red-line positions, revealing frame-to-frame variations over time. Prominent banding, abrupt color shifts, or misaligned stripes in this profile indicate poor temporal consistency, while color distributions that more closely match the ground truth suggest higher geometric accuracy. From Fig. 2(a), the proposed TAN exhibits reduced flickering and fewer depth discontinuities, indicating enhanced temporal coherence. Meanwhile, TAN generates depth values closer to the ground truth, demonstrating superior depth accuracy on tissues while other methods suffer from depth errors accumulated over time, also known as depth drift. From Fig. 2(b), TAN exhibits fewer misaligned stripes in the temporal profile, reflecting superior temporal consistency.

Ablation Study. We investigate the effect of key components, TA and $\mathcal{L}_{tr}$, in the proposed TAN. To establish a baseline, we remove temporal adapters and the temporal regularization loss from TAN, resulting in a model that only fine-tunes DAv2 via LoRA and the depth head with L_s. As shown in the Table 2, incorporating either the TA or L_{tr} into the baseline improves both temporal consistency and depth accuracy, emphasizing the importance of temporal information modeling and temporal regularization in enhancing the stability and accuracy of depth estimation. The inclusion of TA contributes more significantly to improving accuracy, while L_s primarily enhances temporal consistency. Notably, adding four temporal layers to the baseline introduces only 2.8M additional parameters, indicating the parameter efficiency of TA. Furthermore, TA and $\mathcal{L}_{tr}$ complement each other, leading to the best performance. The total trainable parameters (3.1M) of our TAN account for only 3.2% of the total parameters in DAv2 (97.5M), yet it enables higher temporal consistency and depth accuracy, demonstrating its effectiveness for adapting DAv2 from image-based to temporal-aware depth estimation. DAv2 achieves 106 FPS and TAN reaches 97 FPS at a resolution of 224 × 280, highlighting its potential for real-time surgical applications.

4 Conclusion

In this work, we propose a novel TAN for surgical MDE that efficiently incorporates temporal information in the DAv2 foundation model, reducing the inherent instability observed in existing methods. A lightweight temporal adapter is proposed to explicitly model temporal information between frames. A temporal regulation loss is also built to constrain depth variations over time. Extensive experiments on two public datasets demonstrate that TAN outperforms state-of-the-art MDE methods in both temporal consistency and depth accuracy, achieving OPW and RMSE reductions of 14.29–62.42% and 3.6–31.65%, respectively, on the SCARED dataset, and 6.28–45.78% and 3.26–40.86%, respectively, on the EndoNeRF dataset, while maintaining computational efficiency (3.1M trainable parameters, inference at 97 FPS). Qualitative analyses further substantiate the effectiveness of our proposed method in reducing temporally inconsistent artifacts and enhancing depth stability, making it highly suitable for applications requiring consistent depth predictions over time, such as surgical autonomous navigation. Future work will investigate efficient temporal adapters to model longer temporal information and utilize multiple complementary foundation models to mitigate inaccuracies propagated from the optical-flow network.

Acknowledgments. This work was supported in whole, or in part, by the Digital Surgery Ltd, Medtronic; the EPSRC under the UCL Centre for Doctoral Training in Intelligent, Integrated Imaging in Healthcare (i4health) [EP/S021930/1] and the Human-centred Machine Intelligence to optimise Robotic Surgical Training (HuMIRoS) project [EP/Z534754/1]; a UCL Research Excellence Scholarship; the Department of Science, Innovation and Technology (DSIT) and the Royal Academy of Engineering under the Chair in Emerging Technologies programme. For the purpose of open access, the author has applied a CC BY public copyright licence to any author accepted manuscript version arising from this submission.

Disclosure of Interests. The authors have no competing interests to declare that are relevant to the content of this article.

References

1. Alexey, D.: An image is worth 16x16 words: transformers for image recognition at scale. arXiv preprint arXiv: 2010.11929 (2020)
2. Allan, M., et al.: Stereo correspondence and reconstruction of endoscopic data challenge. arXiv preprint arXiv:2101.01133 (2021)
3. Cao, Y., Li, Y., Zhang, H., Ren, C., Liu, Y.: Learning structure affinity for video depth estimation. In: Proceedings of the 29th ACM International Conference on Multimedia, pp. 190–198 (2021)
4. Chen, C., et al.: MA-SAM: modality-agnostic SAM adaptation for 3D medical image segmentation. Med. Image Anal. **98**, 103310 (2024)
5. Chen, L., Tang, W., John, N.W., Wan, T.R., Zhang, J.J.: Slam-based dense surface reconstruction in monocular minimally invasive surgery and its application to augmented reality. Comput. Methods Programs Biomed. **158**, 135–146 (2018)

6. Collins, T., et al.: Augmented reality guided laparoscopic surgery of the uterus. IEEE Trans. Med. Imaging **40**(1), 371–380 (2020)
7. Cui, B., Islam, M., Bai, L., Ren, H.: Surgical-DINO: adapter learning of foundation models for depth estimation in endoscopic surgery. Int. J. Comput. Assisted Radiol. Surg., 1–8 (2024)
8. Cui, B., Islam, M., Bai, L., Wang, A., Ren, H.: EndoDAC: efficient adapting foundation model for self-supervised depth estimation from any endoscopic camera. In: International Conference on Medical Image Computing and Computer-Assisted Intervention, pp. 208–218 (2024)
9. Eigen, D., Puhrsch, C., Fergus, R.: Depth map prediction from a single image using a multi-scale deep network. In: Advances in Neural Information Processing Systems, vol. 27 (2014)
10. Hendrycks, D., Gimpel, K.: Gaussian error linear units (GELUs). arXiv preprint arXiv:1606.08415 (2016)
11. Hu, E.J., et al.: LoRA: low-rank adaptation of large language models. In: International Conference on Learning Representations (2022)
12. Khan, N., Penner, E., Lanman, D., Xiao, L.: Temporally consistent online depth estimation using point-based fusion. In: Proceedings of the IEEE/CVF Conference on Computer Vision and Pattern Recognition, pp. 9119–9129 (2023)
13. Kirillov, A., et al.: Segment anything. In: Proceedings of the IEEE/CVF International Conference on Computer Vision, pp. 4015–4026 (2023)
14. Leonard, S., et al.: Evaluation and stability analysis of video-based navigation system for functional endoscopic sinus surgery on in vivo clinical data. IEEE Trans. Med. Imaging **37**(10), 2185–2195 (2018)
15. Li, B., Liu, B., Zhu, M., Luo, X., Zhou, F.: Image intrinsic-based unsupervised monocular depth estimation in endoscopy. IEEE J. Biomed. Health Inf., 1–11 (2024)
16. Li, Z., et al.: Revisiting stereo depth estimation from a sequence-to-sequence perspective with transformers. In: Proceedings of the IEEE/CVF International Conference on Computer Vision, pp. 6197–6206 (2021)
17. Mahmood, F., Durr, N.J.: Deep learning and conditional random fields-based depth estimation and topographical reconstruction from conventional endoscopy. Med. Image Anal. **48**, 230–243 (2018)
18. Oquab, M., et al.: DINOv2: learning robust visual features without supervision. arXiv preprint arXiv:2304.07193 (2023)
19. Ranftl, R., Bochkovskiy, A., Koltun, V.: Vision transformers for dense prediction. In: Proceedings of the IEEE/CVF International Conference on Computer Vision, pp. 12159–12168 (2021)
20. Shao, S.: Self-supervised monocular depth and ego-motion estimation in endoscopy: appearance flow to the rescue. Med. Image Anal. **77**, 102338 (2022)
21. Wang, Y., Pan, Z., Li, X., Cao, Z., Xian, K., Zhang, J.: Less is more: consistent video depth estimation with masked frames modeling. In: Proceedings of the 30th ACM International Conference on Multimedia, pp. 6347–6358 (2022)
22. Wang, Y., et al.: NVDS$^+$+: towards efficient and versatile neural stabilizer for video depth estimation. IEEE Trans. Pattern Anal. Mach. Intell. **47**(1), 583–600 (2025)
23. Wang, Y., Long, Y., Fan, S.H., Dou, Q.: Neural rendering for stereo 3D reconstruction of deformable tissues in robotic surgery. In: International Conference on Medical Image Computing and Computer-Assisted Intervention, pp. 431–441 (2022)

24. Xu, H., Zhang, J., Cai, J., Rezatofighi, H., Tao, D.: GMFlow: learning optical flow via global matching. In: Proceedings of the IEEE/CVF Conference on Computer Vision and Pattern Recognition, pp. 8121–8130 (2022)
25. Yang, L., Kang, B., Huang, Z., Xu, X., Feng, J., Zhao, H.: Depth anything: unleashing the power of large-scale unlabeled data. In: Proceedings of the IEEE/CVF Conference on Computer Vision and Pattern Recognition, pp. 10371–10381 (2024)
26. Yang, L., et al.: Depth anything V2. In: The Thirty-Eighth Annual Conference on Neural Information Processing Systems (2024)

Real-Time Surgical Keypoint Detection in Laparoscopic Cholecystectomy

Yiyang You[1], De Ru Tsai[1(✉)], Yoseph Kim[1], Antony Goldenberg[2], Juo Tung Chen[2], Ji Woong Brian Kim[3], Axel Krieger[2], and Richard Jaepyeong Cha[4]

[1] Optosurgical, LLC, Columbia, MD 21046, USA
deru.tsai@optosurgical.com
[2] Laboratory for Computational Sensing and Robotics, Johns Hopkins University, Baltimore, MD 21218, USA
[3] Stanford Artificial Intelligence Laboratory, Stanford University, Stanford, CA 94305, USA
[4] Sheikh Zayed Institute for Pediatric Surgical Innovation, Children's National Hospital, Washington, DC 20010, USA

Abstract. Accurate intraoperative tracking of keypoints is vital for trajectory planning in autonomous robotic surgery. Traditional methods for keypoint detection often rely on heatmap-based techniques, which can be computationally intensive and may not seamlessly integrate into real-time surgical workflows. In this study, we introduce a deep learning-based method for keypoint detection on soft tissue, modifying and optimizing YOLO-v8n for efficient localization of grasping and clipping points during LC using ex vivo porcine gallbladder tissues. Distinct from YOLO-v8n-pose, which focuses on learning keypoint heads, our model is trained to produce bounding boxes with their centers corresponding to the keypoints. Our approach has demonstrated an accuracy of 0.886, defined as the proportion of predicted points that fall within a defined range of the ground truth while achieving a performance of around 120 frames per second (FPS). It also outperformed previous keypoint detection models and achieved a notable reduction in computational time. These results endorse the feasibility of integration into surgical guidance systems for enhancing minimally invasive procedures in real time.

Keywords: Laparoscopic cholecystectomy · keypoint detection · real-time tissue tracking · surgical guidance system

1 Introduction

Bile duct injury (BDI), a serious complication of laparoscopic cholecystectomy (LC), occurs in 0.3% to 0.7% of cases and poses significant risks. With over

---

Y. You and D. R. Tsai—These authors contributed equally to this work.

750,000 LCs performed annually in the US [1], preventing BDI remains challenging due to the complexity of the biliary system. Accurate anatomical identification in robotic-assisted LC has been shown to improve outcomes, such as shorter hospital stays and lower readmission rates [2]. While recent studies have applied AI to structure recognition and decision support during LC [3–5], limited work has focused on detecting keypoints related to critical actions like grasping, clipping, and cutting, which are vital for precise trajectory planning in autonomous surgery. To address this issue, we propose a modified YOLO-v8n framework to enhance real-time keypoint detection in deformable soft tissue during minimally invasive procedures.

1.1 Related Works

Keypoint detection for surgical guidance and robotic-assisted procedures remains an active area of research due to the variable anatomical features. A previous study, utilizing a CNN model to predict keypoint heatmaps for knee radiographs [6], motivates our exploration of intensity generation that reflects the probability of the keypoint being located at that position. Although many prominent studies predominantly apply this approach to human pose estimation, we can still leverage their heatmap generation components for surgical keypoint detection.

In the realm of keypoint detection, heatmap-based methods have gained prominence, with SimpleBaseline [7] and HRNet [8] emerging as notable benchmarks. SimpleBaseline is characterized by an architecture that employs ResNet [9] as the backbone for feature extraction and subsequently transforms the extracted features into heatmaps. HRNet enhances this approach by maintaining multiple parallel resolutions of features throughout the network, thereby avoiding the need for progressive downsampling. However, a common limitation of these methods is their reliance on explicit heatmap generation, which can incur substantial computational overhead. This limitation draws our attention to the CenterNet framework [10]. Although CenterNet is not traditionally employed for keypoint detection, it predicts object centers on a heatmap while simultaneously regressing their sizes and offsets to facilitate object detection. While deep architecture frameworks such as Hourglass [11] or ResNet introduce considerable computational demands, their strong representational power suggests the potential transfer from object detection to keypoint detection.

1.2 YOLO vs. Prior Works

Among various object detection models [12–14], YOLO [15,16] stands out for its efficiency, dividing an image into a grid and predicting bounding boxes and class probabilities in a single forward pass. While traditional YOLO models do not generate heatmaps for keypoints, a modified version incorporates keypoint prediction as an additional output [17]. This modification utilizes the centers of the predicted bounding boxes as anchors for keypoint predictions associated with the detected objects. Compared to heatmap-based methods, this design requires fewer parameters, as it downsamples the image into grids rather than processing the entire image through all stages. It also eliminates the need to generate numerous heatmaps for all keypoints. This approach raises an intriguing question: Is there a more straightforward way to predict keypoints based on the bounding boxes? This idea prompts us to investigate the feasibility of training a YOLO model that uses the centers of the bounding boxes directly as keypoints.

1.3 Problem Definition

In this study, we focus on the clipping and cutting processes in LC. Once the critical view of safety (CVS) is established, the tasks of clip placement and duct/artery transaction can be precisely modeled. The procedure proceeds as follows: the grasper applies tension to the neck of the gallbladder, stretching the tubes apart to create sufficient space for tools to access the gap. Subsequently, three clips are placed on the left structure (typically the cystic duct), and another three clips on the right structure (typically the cystic artery). Transection is performed between the second and third clips, where the gap is most accessible for scissor insertion.

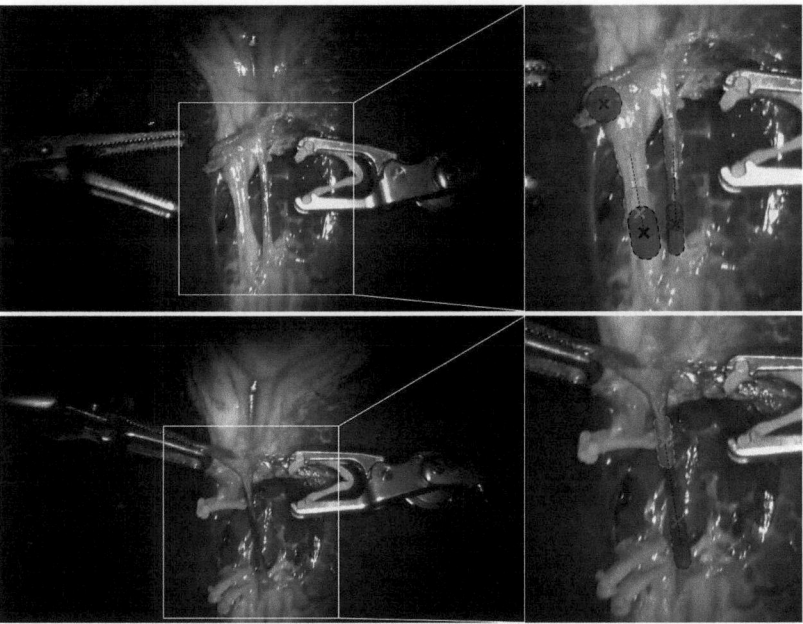

Fig. 1. Predictions falling within a defined range around the ground truth (indicated by cross marks) are considered accurate. Keypoints occluded by instruments or clips are excluded. (Green: grasping; Blue: duct clipping; Red: artery clipping) (Color figure online)

To model the keypoint detection process, we define seven keypoints guided by the operative steps after the demonstrators establish CVS. These include one grasping point on the gallbladder, and six clipping points: three on the cystic duct—two at the proximal end and one at the distal end near the gallbladder—and three on the cystic artery, arranged in a similar configuration. Keypoints that are occluded by surgical instruments or clips are excluded to prevent the model from misidentifying irrelevant features. Depending on the surgical phase, the model may predict up to seven keypoints (during the initial dissection) and as few as zero (once clipping and cutting are complete).

Since point locations may vary slightly among demonstrators due to individual preferences, we consider predictions within a defined tolerance range to be acceptable. As shown in Fig. 1, the range of a grasping point is defined as 20% of the gallbladder head width, and the grasping points are evaluated using a centerline drawn along the tube structure, extending from the first to the third point. Based on this centerline, the distance between the predicted and ground truth points is decomposed into vertical and horizontal components. A prediction is considered correct if the vertical distance is less than half the tissue width and the horizontal distance is less than 15% of the tissue length. Predictions falling outside these ranges or incorrectly classified are considered negative.

2 Method

2.1 Dataset

Prior to data collection, a surgeon performed blunt dissection of normal porcine gallbladders to establish CVS. Two experienced demonstrators then performed key surgical steps, including grasping the gallbladder head, applying clips to the cystic duct and artery, and performing dissection. Visual data were recorded using the dVRK stereo endoscope at a resolution of 960×540 pixels, with one frame per second extracted. The training set comprised 2,000 images from 20 gallbladders under various surgical conditions, including dissection and clip placement. YOLO's default data augmentation techniques (random horizontal flipping, affine transformations, HSV color augmentation, and mosaic augmentation) were applied during training. The testing set comprised 200 images from 5 unseen gallbladders, following the same distribution as the training set.

The annotation of the keypoints was guided by the operative steps, ensuring alignment with real surgical practice. Since all seven keypoint-related actions were performed during each procedure, the complete surgical videos enabled precise identification of the corresponding anatomical locations on each tissue. This ensured both the spatial accuracy and the temporal consistency of the keypoint labels across frames and subjects.

2.2 Model

Rather than employing conventional heatmap-based keypoint detection, we adopt YOLO-v8n, a real-time object detection framework optimized for dynamic tissue tracking. We chose version 8 since it benefits from a more mature ecosystem compared to versions 9 and 10 [18,19], alongside a simpler architectural design compared to Version 11 [20], which facilitates modifications.

Given N keypoints, YOLO-v8n outputs predictions for each grid cell as:

$$P = \{C_x, C_y, W, H, Box_{conf}, Class^1_{conf}, \ldots, Class^N_{conf}\} \quad (1)$$

Here, C_x and C_y denote the center coordinate, W and H denote the predicted bounding box dimensions, Box_{conf} denotes the box confidence score, and $Class^N_{conf}$ denotes the confidence score for class N. In our implementation, we use C_x and C_y directly as the keypoint coordinates. This results in a more compact output representation compared to YOLO-v8 pose, which predicts:

$$P = \{C_x, C_y, W, H, Box_{conf}, Class_{conf}, K^1_x, K^1_y, \ldots, K^N_x, K^N_y\} \quad (2)$$

Here, K^N_x and K^N_y denote the coordinates of the N-th keypoint. This streamlined output reduces computational complexity and simplifies keypoint representation.

Our first architectural modification replaces the Convolutional to Fused (C2f) module with a C3STR block, which incorporates the Swin Transformer (SwinT) [21] into the cross-stage partial (CSP) framework. While the original C2f module effectively fuses local features via convolution, it lacks the capacity to model long-range dependencies and global anatomical context. C3STR addresses this by splitting the input into two branches: one passes through a SwinT block using window-based multi-head self-attention with shifted windows, and the other bypasses it. The fused output enables the network to jointly capture local details (e.g., tools, clips) and broader anatomical structures (e.g., CVS) [22].

We also replace the Spatial Pyramid Pooling Fast (SPPF) layer with the Spatial Pyramid Pooling FastâĂŞCross Stage Partial Connections (SPPF-CSPC) layer [23]. While SPPF aggregates multi-scale features via sequential pooling, it suffers from redundancy and limited information flow. SPPF-CSPC mitigates these issues by splitting the input feature map, processing part through the pooling path, and merging it with bypassed features. This design preserves fine-grained spatial details and enhances gradient propagation, benefiting the detection of small, variable targets.

The keypoints were used as centers for bounding box labels. We initially employed smaller bounding boxes to tightly focus on the target structures. However, excessively small boxes risk omitting essential contextual information. To assess this trade-off, we evaluated model performance across varying box sizes.

3 Experiment and Results

3.1 Implementation Details

All experiments were conducted on an NVIDIA RTX^{TM} 6000 Ada Generation 48 GB GPU. For training the YOLO-v8n models, we employed AdamW with an initial learning rate of 9.09e−4. Training was performed for 200 epochs with a batch size of 32 and a weight decay of 5.0e−4. For the other selected models, we used their default settings with the same number of epochs and batch size as the YOLO-v8n models. No pretrained weights were used for any model.

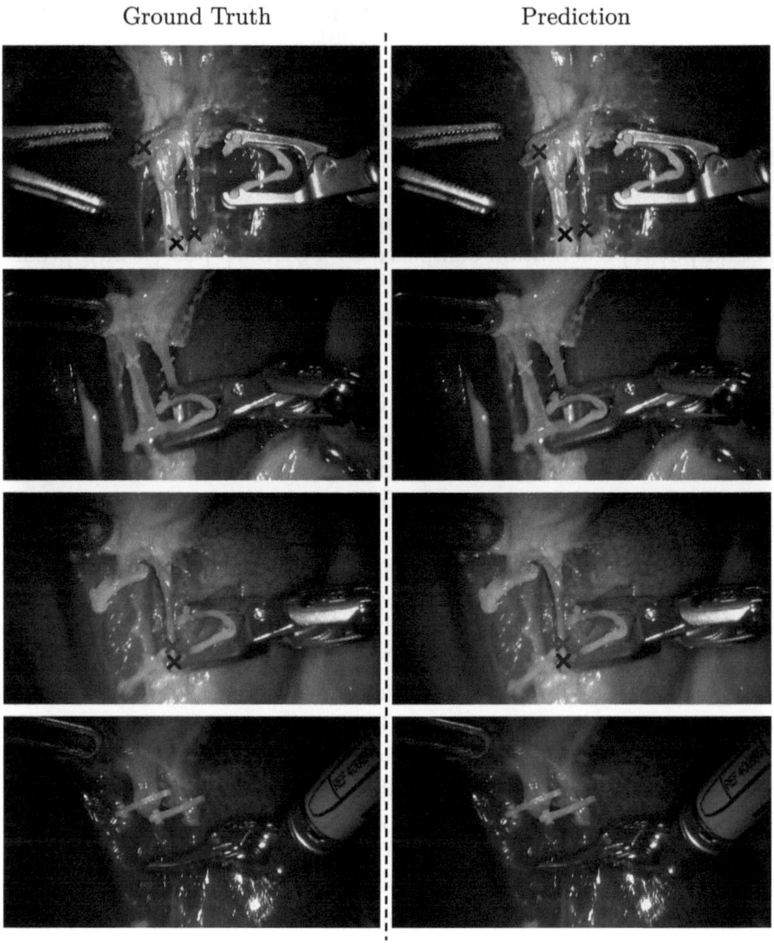

Fig. 2. The results of the ground truth (left) and the prediction (right) on an unseen gallbladder over multiple steps highlight its ability to predict keypoints. The images presented here were cropped to provide better visualization.

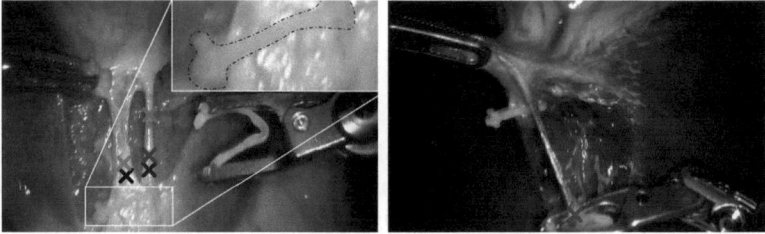

Fig. 3. The model may misclassify clips closely resembling surroundings to tissue (left) or fail to recognize deformed tissues (right).

Table 1. Comparison using our ex vivo porcine gallbladder testing set.

Method	Backbone	Params #	Input size	FPS	ACC	PREC.	Recall	F1-score
SimpleBaseline	ResNet-50	34.1M	384 × 640	9.52	0.5440	0.3314	0.3176	0.3243
SimpleBaseline	ResNet-101	53.1M	384 × 640	7.78	0.5768	0.3462	0.4684	0.3981
SimpleBaseline	ResNet-152	68.8M	384 × 640	6.86	0.6212	0.4028	0.7447	0.5228
HRNet-W32	HRNet-W32	28.5M	384 × 640	3.59	0.7814	0.7060	0.8556	0.7736
HRNet-W48	HRNet-W48	63.6M	384 × 640	3.33	0.7468	0.6393	0.8342	0.7238
YOLO-v8n-pose	CSPNet	4.9M	384 × 640	164	0.8005	0.6161	0.9011	0.7317
YOLO-v8n-original (center as keypoint)	CSPNet	4.8M	384 × 640	161	0.8603	0.8348	0.8251	0.8299
YOLO-v8n-modified (center as keypoint)	CSPNet + SwinT	6.3M	384 × 640	123	**0.8860**	**0.8349**	**0.9000**	**0.8662**

3.2 Results

We compared our method against established benchmarks, SimpleBaseline and HRNet, using the COCO pose format, where each keypoint includes a coordinate and visibility. All annotated keypoints were assigned a visibility value of 2. Optimal performance was achieved with a visibility threshold of 0.7 for SimpleBaseline and 0.8 for HRNet. As demonstrated in Fig. 1, our method accurately detects keypoints in most scenarios. Table 1 shows that it achieves competitive performance with significantly fewer model parameters and shorter inference time, and outperforms all previous methods in all other metrics.

In experiments with the YOLO-v8n-pose, we provided additional bounding box labels around key anatomical regions containing the seven defined keypoints to help the model focus on relevant context. Although YOLO-v8n-pose achieves a strong recall (0.9011), its lower precision (0.6161) suggests that it may have learned ambiguous patterns. Additionally, some predictions even fall outside the target organs, further reducing overall accuracy and precision.

Table 2. Comparison on different scenarios.

Step	ACC	PREC.	Recall	F1-score
Grasping gallbladder	0.8000	0.8559	0.9076	0.8810
Clipping second clip on the left tube	0.9007	0.8503	0.9470	0.8961
Clipping first clip on the right tube	0.8407	0.7264	0.7549	0.7404
After transecting	1.0000	0.0000	0.0000	0.0000

Compared to the original YOLO-v8n model, our modification improved accuracy by 2.57%, recall by 7.49%, and F1-score by 3.63%. In our case, the optimal

Table 3. Distance between ground truth and predictions from our modified model. Distance^V stands for the vertical distance, and Distance^H stands for the horizontal distance according to the center line of the cystic duct or artery.

Point type	Distance/Gallbladder head width (%)	
Gallbladder grasping	16.14 ± 3.46	
Point type	Distance^V/Tissue width (%)	Distance^H/Tissue length (%)
Duct clipping[1]	30.98 ± 9.01	9.19 ± 3.60
Duct clipping[2]	15.57 ± 10.48	6.49 ± 3.49
Duct clipping[3]	10.89 ± 6.31	5.82 ± 3.11
Artery clipping[1]	17.32 ± 19.03	6.62 ± 3.78
Artery clipping[2]	16.27 ± 18.59	6.27 ± 4.31
Artery clipping[3]	13.78 ± 12.92	6.20 ± 4.51

normalized bounding box size for labels was 0.12 × 0.10. Fig. 2 shows keypoint prediction across four representative surgical steps: (1) before grasping the gallbladder, (2) after placing the first clip on the left structure, (3) after transecting the left structure, and (4) after completing all procedures. Table 2 confirms strong alignment with ground truth across all steps. As shown in Table 3, all predicted keypoints fall within the defined acceptable error range, indicating reliable presence detection and precise localization. The model also performs robustly under occlusion and after completed actions, demonstrating strong spatial awareness and temporal consistency.

4 Discussion and Summary

Our method accurately detects keypoints across multiple surgical steps and outperforms previous models, but remains sensitive to occlusions and instances where clips closely resemble surrounding tissue. Instrument interference and varying tissue pose can also impair recognition. As shown in Fig. 3, the left image shows a failure case where the model incorrectly predicts a keypoint for a completed action due to a clip resembling nearby tissue. The right image shows missed keypoints caused by tissue deformation, specular reflections, and a limited field of view. Unlike human pose estimation models, our approach has limitations in keypoint grouping and does not assess anatomical alignment. Additionally, the absence of anomalies in the training data restricts prediction in such cases. To improve generalization, we plan to integrate attention mechanisms for enhanced global context processing. Our results also highlight the importance of selecting appropriate bounding box sizes during label post-processing. Next, we aim to incorporate autonomous bounding box size selection into the learning process.

Although the demonstration did not cover the entire LC procedure, and annotations may vary between demonstrators, this study presents a generalized framework for training models to predict keypoints in medical images. By training separate models for individual surgical steps, our approach has the potential to enhance accuracy, especially when combined with a surgical step identification model. We believe this strategy could support real-time interaction with surgeons or robotic systems during minimally invasive surgery, ultimately contributing to improved surgical outcomes (Table 4).

Table 4. Comparison between bounding box sizes of labels.

Normalized bounding box size	ACC	PREC.	Recall	F1-score
0.08 × 0.06	0.8596	0.8537	0.8063	0.8293
0.10 × 0.08	0.8603	0.8358	0.8297	0.8327
0.12 × 0.10	**0.8860**	**0.8349**	**0.9000**	**0.8662**
0.14 × 0.12	0.8801	0.8542	0.8625	0.8583

References

1. Pesce, A., et al.: Iatrogenic bile duct injury: impact and management challenges. Clin. Exp. Gastroenterol. **12**, 121–128 (2019)
2. Kane, W.J., et al.: Robotic compared with laparoscopic cholecystectomy: a propensity matched analysis. Surgery **167**(2), 432–435 (2020)
3. Fernicola, A., et al.: Artificial intelligence applied to laparoscopic cholecystectomy: what is the next step? A narrative review. Updates Surg. **76**(5), 1655–1667 (2024)
4. Orimoto, H., et al.: Development of an artificial intelligence system to indicate intraoperative findings of scarring in laparoscopic cholecystectomy for cholecystitis. Surg. Endos. 1–9 (2025)
5. Corallino, D., et al.: Systematic review on the use of artificial intelligence to identify anatomical structures during laparoscopic cholecystectomy: a tool towards the future. Langenbecks Arch. Surg. **410**(1), 1–16 (2025)
6. Nam, H.S., et al.: Key-point detection algorithm of deep learning can predict lower limb alignment with simple knee radiographs. J. Clin. Med. **12**(4), 1455 (2023)
7. Xiao, B., Wu, H., Wei, Y.: Simple baselines for human pose estimation and tracking. In: Ferrari, V., Hebert, M., Sminchisescu, C., Weiss, Y. (eds.) ECCV 2018. LNCS, vol. 11210, pp. 472–487. Springer, Cham (2018). https://doi.org/10.1007/978-3-030-01231-1_29
8. Sun, K., et al.: Deep high-resolution representation learning for human pose estimation. In: Proceedings of the IEEE/CVF Conference on Computer Vision and Pattern Recognition, pp. 5693–5703 (2019)
9. He, K., et al.: Deep residual learning for image recognition. In: Proceedings of the IEEE Conference on Computer Vision and Pattern Recognition, pp. 770–778 (2016)
10. Duan, K., et al.: Centernet: keypoint triplets for object detection. In: Proceedings of the IEEE/CVF International Conference on Computer Vision, pp. 6569–6578 (2019)
11. Newell, A., Yang, K., Deng, J.: Stacked hourglass networks for human pose estimation. In: Leibe, B., Matas, J., Sebe, N., Welling, M. (eds.) ECCV 2016. LNCS, vol. 9912, pp. 483–499. Springer, Cham (2016). https://doi.org/10.1007/978-3-319-46484-8_29

12. Ren, S., et al.: Faster R-CNN: towards real-time object detection with region proposal networks. In: Advances in Neural Information Processing Systems, vol. 28 (2015)
13. Liu, W., et al.: SSD: single shot multibox detector. In: Leibe, B., Matas, J., Sebe, N., Welling, M. (eds.) ECCV 2016, Part I. LNCS, vol. 9905, pp. 21–37. Springer, Cham (2016). https://doi.org/10.1007/978-3-319-46448-0_2
14. Carion, N., Massa, F., Synnaeve, G., Usunier, N., Kirillov, A., Zagoruyko, S.: End-to-end object detection with transformers. In: Vedaldi, A., Bischof, H., Brox, T., Frahm, J.-M. (eds.) ECCV 2020. LNCS, vol. 12346, pp. 213–229. Springer, Cham (2020). https://doi.org/10.1007/978-3-030-58452-8_13
15. Jocher, G., et al.: Ultralytics YOLO. https://github.com/ultralytics/ultralytics. Accessed 14 July 2024
16. Hussain, M., et al.: YOLO-v1 to YOLO-v8, the rise of YOLO and its complementary nature toward digital manufacturing and industrial defect detection. Machines **11**(7), 677 (2023)
17. Maji, D., et al.: YOLO-pose: enhancing yolo for multi person pose estimation using object keypoint similarity loss. In: Proceedings of the IEEE/CVF Conference on Computer Vision and Pattern Recognition, pp. 2637–2646 (2022)
18. Wang, C., et al.: YOLOv9: learning what you want to learn using programmable gradient information. In: Leonardis, A., Ricci, E., Roth, S., Russakovsky, O., Sattler, T., Varol, G. (eds.) ECCV 2024. LNCS, vol. 15089, pp. 1–21. Springer, Cham (2024). https://doi.org/10.1007/978-3-031-72751-1_1
19. Wang, A., et al.: YOLOv10: real-time end-to-end object detection. In: Advances in Neural Information Processing Systems, vol. 37, pp. 107984–108011 (2025)
20. Khanam, R., et al.: YOLOv11: an overview of the key architectural enhancements. arXiv preprint arXiv:2410.17725 (2024)
21. Liu, Z., et al.: Swin transformer: hierarchical vision transformer using shifted windows. In: Proceedings of the IEEE/CVF International Conference on Computer Vision (ICCV), pp. 10012–10022 (2021)
22. Kumar, A., et al.: An improved feature extraction algorithm for robust Swin Transformer model in high-dimensional medical image analysis. Comput. Biol. Med. **188**, 109822 (2025)
23. Li, C., et al.: YOLOv6 v3.0: a full-scale reloading. arXiv preprint arXiv:2301.05586 (2023)

When Tracking Fails: Analyzing Failure Modes of SAM2 for Point-Based Tracking in Surgical Videos

Woowon Jang[1], Jiwon Im[1], Juseung Choi[1], Niki Rashidian[2,3], Wesley De Neve[1,4], and Utku Ozbulak[1,4(✉)]

[1] Center for Biosystems and Biotech Data Science, Ghent University Global Campus, Incheon, Republic of Korea
[2] Department of Human Structure and Repair, Ghent University, Ghent, Belgium
[3] Department of HPB Surgery and Liver Transplantation, Ghent University Hospital, Ghent, Belgium
[4] IDLab, Department of Electronics and Information Systems, Ghent University, Ghent, Belgium
utku.ozbulak@ghent.ac.kr

Abstract. Video object segmentation (VOS) models such as SAM2 offer promising zero-shot tracking capabilities for surgical videos using minimal user input. Among the available input types, point-based tracking offers an efficient and low-cost alternative, yet its reliability and failure cases in complex surgical environments are not well understood. In this work, we systematically analyze the failure modes of point-based tracking in laparoscopic cholecystectomy videos. Focusing on three surgical targets, the gallbladder, grasper, and L-hook electrocautery, we compare the performance of point-based tracking with segmentation mask initialization. Our results show that point-based tracking is competitive for surgical tools but consistently underperforms for anatomical targets, where tissue similarity and ambiguous boundaries lead to failure. Through qualitative analysis, we reveal key factors influencing tracking outcomes and provide several actionable recommendations for selecting and placing tracking points to improve performance in surgical video analysis.

Keywords: Video object segmentation · SAM2 · Surgical video understanding · Surgical AI

1 Introduction

Artificial intelligence (AI)-based technologies are increasingly being integrated into surgical workflows to enhance intraoperative decision-making, postoperative analysis, and practitioner training [12,14]. Among these, AI-based surgical video understanding has gained particular attention due to its potential to

---

W. Jang, J. Im, J. Choi—Equal contribution.

automate annotation, identify anatomical structures, and provide context-aware guidance during surgical procedures [15]. The ability to interpret visual data from robotic surgeries offers promising avenues for improving surgical safety, efficiency, and documentation [2]. However, building reliable AI systems for such high-stakes environments remains a considerable challenge due to the complex and dynamic nature of surgical scenes [9].

Video object segmentation models, particularly those based on vision foundation models such as Segment Anything Model 2 (SAM2), have emerged as promising tools for surgical video understanding. These models can propagate object-level information across long video sequences while maintaining spatial and temporal consistency [4]. In particular, SAM2 and several of its medical variants such as MedSAM2 and Surgical SAM2 leverage transformer-based attention mechanisms to perform zero-shot segmentation and tracking across frames, making it highly adaptable to the medical field where labeled data are limited [5]. In the context of surgery, such models could support applications ranging from instrument tracking and anatomy localization to retrospective analysis of procedural steps [7].

Tracking performance in video object segmentation heavily depends on the type of user input used to initiate and guide object propagation. Tracking can be performed in three ways: (1) via full segmentation masks, (2) via bounding boxes, and (3) via tracking points [18]. Among these, segmentation masks offer the most accurate object representation but are time-consuming and expensive to generate, especially in surgical videos where an expert is needed for the annotation [10,11]. Bounding boxes, while more convenient, are often ill-suited for fine-grained surgical tracking due to their coarse spatial localization. Point-based tracking, which involves selecting a few reference points on the object of interest, offers an efficient and low-cost alternative. Several recent studies have suggested that, under ideal conditions, point-based tracking has the potential to reach segmentation-level accuracy, making it an attractive target for real-world surgical applications [17]. Despite its practical appeal, the reliability and failure modes of point-based tracking in complex surgical scenarios remain poorly understood. In this work, we tackle this gap by conducting a systematic investigation into the performance and failure modes of point-based tracking in complex surgical environments. To that end, we systematically investigate the failure modes of SAM2 when used for point-based object tracking in surgical videos, focusing on laparoscopic cholecystectomy.

Our results show that point-based tracking is almost always inferior to segmentation mask-based tracking, particularly for anatomical targets. Notably, we observe that point-based tracking fails in distinct ways for the gallbladder, exhibiting issues such as shape misalignment and failure to adapt to anatomical deformation. In contrast, for surgical instruments such as the grasper and L-hook electrocautery, point-based tracking performs considerably better, sometimes approaching the accuracy of segmentation mask-based tracking. Finally, we present a range of qualitative examples highlighting these failure modes, providing practical insights into when and why point-based tracking succeeds or fails in surgical video analysis.

2 Methodology

2.1 Model

Recent advances in VOS models, particularly transformer-based foundation models, have shown strong zero-shot performance in complex domains such as surgical video analysis [5,9]. In this work, we leverage SAM2.1 Hiera Large [18], a state-of-the-art zero-shot segmentation model that employs a hierarchical transformer architecture for robust spatial and temporal reasoning. SAM2 has shown remarkable performance in medical image segmentation and surgical video understanding [3,16,19], thanks in part to its built-in tracking mechanism that reduces frame-wise segmentation drift and ensures stable object propagation over time. This temporal consistency is particularly important for surgical video analysis, where maintaining coherent tracking of instruments and anatomical structures supports real-time decision-making and situational awareness in the operating room [12].

Several adaptations of SAM2 have been proposed for medical applications, including MedSAM2 for static medical images and Surgical-SAM2 for video-based surgical tracking [11,13]. Surgical-SAM2 improves computational efficiency through frame pruning and iterative refinement, making it well suited for offline or sparse-frame video processing [11]. However, in this study, we intentionally avoid using Surgical-SAM2 to preserve the native 25 frames per second (FPS) video streams without pruning, allowing us to assess the tracking performance of SAM2 in continuous, real-time surgical video streams.

2.2 Data

For this study, we use a subset of the CholecSeg8k dataset [6], a frame-level annotated surgical video dataset widely used for evaluating segmentation and tracking models in laparoscopic procedures. While CholecSeg8k contains a broader set of annotated videos and object categories, we specifically select ten video segments and three target objects to conduct a focused and detailed analysis of tracking performance. This subset strikes a balance between covering sufficient surgical variability and allowing for a thorough qualitative and quantitative study of failure cases.

We focus our analysis on three key targets encountered during cholecystectomy procedures: the gallbladder, the grasper, and the L-hook electrocautery instrument. These objects represent two distinct tracking challenges, deformable anatomy and rigid surgical tools, providing two diverse scenarios for evaluating the robustness of point-based tracking. We intentionally exclude additional surgical targets, such as the liver, connective tissue, and fat, to maintain a clear experimental scope and to avoid diluting our analysis across too many object categories. This allows us to concentrate on a representative yet manageable set of tracking scenarios that reveal distinct failure modes.

2.3 Tracking Points

To investigate how the choice and placement of tracking points affect performance, we generate tracking points using three distinct strategies: k-medoids clustering, Shi-Tomasi corner detection, and random sampling.

- **K-medoids**: Points are placed near the geometric centers of the object mask, resembling centroid-based initialization [8].
- **Shi-Tomasi**: Points are placed close to the edges of the object [1].
- **Random**: Points are sampled uniformly from the object mask, simulating arbitrary point selection without prior knowledge.

More details about point selection using k-medoids clustering and Shi-Tomasi corner detection can be found in the work of [17].

For each strategy, we run experiments using 20 different random seeds, allowing us to assess variability due to different point placements. Additionally, we evaluate the impact of the number of tracking points by varying the quantity across five settings: 1, 2, 3, 5, and 7 points.

3 Experimental Results

Using the setup described in Sect. 2, we conduct experiments to evaluate the tracking performance of SAM2 across different objects, videos, and tracking point configurations. Our goal is to compare point-based tracking against segmentation mask-based tracking and analyze how the number and placement of points influence performance.

In Table 1, we present the highest average IoU scores obtained for each object and video segment across all point selection strategies. For each video, we report the best result achieved with 1, 2, 3, 5, or 7 points, allowing us to isolate the upper performance limit of point-based tracking irrespective of initialization method. Our observations from this table are as follows:

- For the gallbladder, segmentation mask-based tracking consistently outperforms point-based tracking across all videos, highlighting the challenges of tracking anatomical structures with tracking points.
- For the grasper and L-hook electrocautery, point-based tracking achieves competitive performance, with several videos showing minimal difference between points and segmentation. This suggests that surgical tools are easier to track with tracking points.
- The gap between segmentation and point-based tracking is larger for anatomical targets than for surgical tools.
- Increasing the number of points generally improves tracking performance, but does not fully close the gap with segmentation masks for anatomical targets. For surgical tools, even a small number of points can yield strong results.

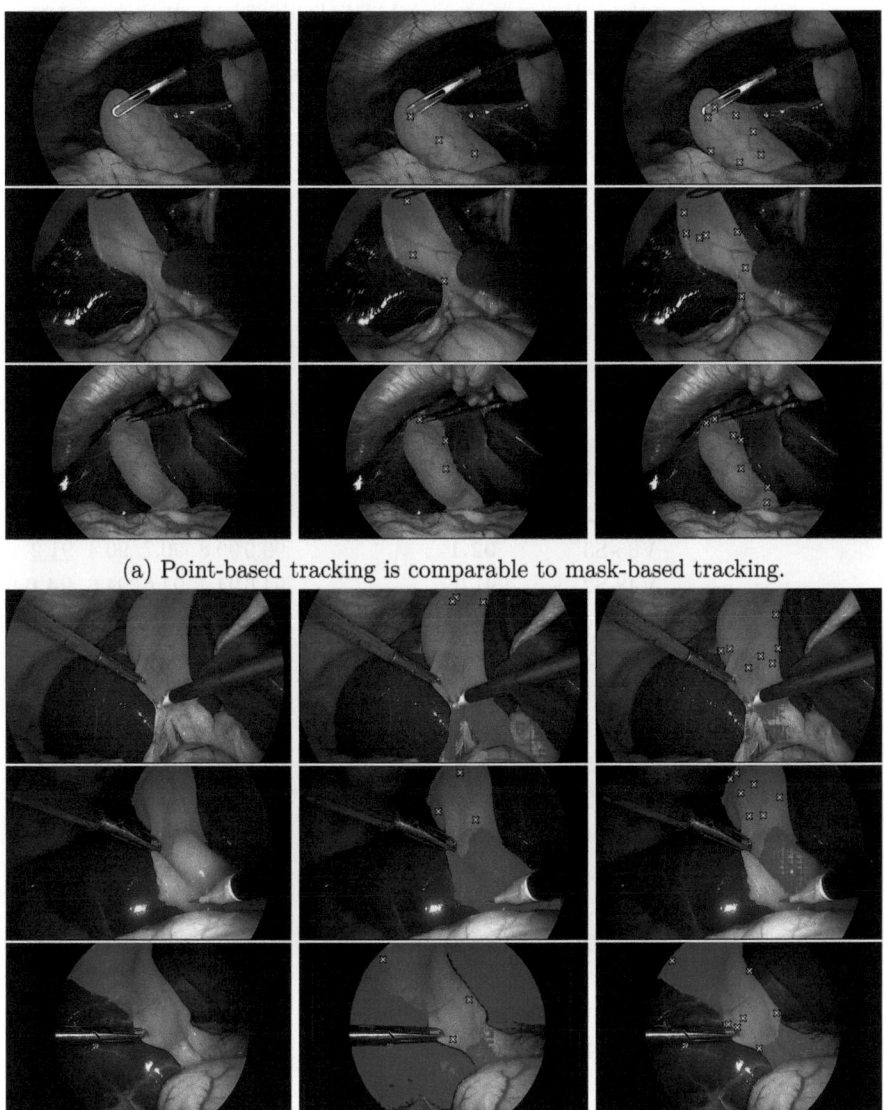

(a) Point-based tracking is comparable to mask-based tracking.

(b) Mask-based tracking is substantially better than point-based tracking.

Fig. 1. Qualitative examples illustrating two distinct tracking outcomes (a) and (b) for the gallbladder. Images in the first column show the ground truth segmentation masks; the second column shows the highest IoU case with 3 tracking points, and the third column shows the highest IoU case with 7 tracking points. Gallbladder regions correctly predicted are highlighted in pink, and incorrect predictions are highlighted in red. Tracking points are shown as white X marks with black outlines. (Color figure online)

Table 1. Highest average IoU scores for each object and video segment across different tracking point configurations. For each row, we highlight the best-performing setup in **bold font** and underline the highest score among the point-based tracking configurations. The *Segmentation Mask* column reports the IoU when the full object mask is used for tracking initialization. Results for 1, 2, 3, 5, and 7 points correspond to the best score obtained across all point selection strategies.

Tracking Target	Video Segment	Segmentation Mask	Number of Tracking Points				
			1 pt	2 pts	3 pts	5 pts	7 pts
Gallbladder	V01-S4	**85.3**	57.8	52.4	52.0	51.7	<u>62.1</u>
	V17-S2	**95.3**	94.3	94.0	94.4	<u>94.9</u>	<u>94.9</u>
	V18-S1	**88.2**	67.4	66.0	60.8	67.7	<u>75.7</u>
	V20-S1	**93.1**	42.4	43.8	43.5	<u>57.1</u>	55.4
	V24-S5	**96.1**	51.3	51.5	55.4	54.1	<u>55.8</u>
	V26-S2	91.4	88.4	91.0	91.2	91.3	**<u>91.7</u>**
	V35-S1	**93.7**	86.9	89.7	89.4	90.0	<u>90.2</u>
	V35-S2	**95.0**	92.9	93.2	93.2	93.5	<u>94.3</u>
	V35-S3	**92.1**	90.9	90.8	90.7	90.4	<u>91.2</u>
	V48-S2	91.6	<u>94.6</u>	94.4	<u>94.6</u>	<u>94.6</u>	**<u>94.6</u>**
Grasper	V01-S4	**81.9**	63.0	65.8	65.6	68.1	<u>68.4</u>
	V17-S2	**86.1**	84.1	84.0	84.2	<u>84.4</u>	84.3
	V18-S1	94.3	94.7	**<u>94.8</u>**	94.7	94.7	94.7
	V20-S1	**94.6**	93.4	93.5	93.6	<u>93.8</u>	<u>93.8</u>
	V24-S5	90.6	90.8	90.9	91.0	**<u>91.2</u>**	91.0
	V26-S2	**77.3**	75.5	<u>75.6</u>	75.4	75.1	75.2
	V35-S1	94.3	94.2	<u>94.3</u>	<u>94.3</u>	<u>94.3</u>	<u>94.3</u>
	V35-S2	**93.1**	92.4	<u>92.4</u>	<u>92.4</u>	<u>92.4</u>	92.3
	V35-S3	**96.2**	95.9	95.9	95.9	95.9	<u>96.0</u>
	V48-S2	**80.1**	76.4	75.7	<u>76.8</u>	75.9	76.6
L-hook Electrocautery	V01-S4	90.8	92.3	**<u>92.6</u>**	89.9	89.9	90.0
	V18-S1	**94.4**	93.4	93.4	<u>93.5</u>	<u>93.5</u>	<u>93.5</u>
	V20-S1	**91.7**	90.9	91.0	90.9	91.0	<u>91.2</u>

We now perform a deeper investigation of the gallbladder tracking results by categorizing the cases into two groups: (1) cases where segmentation mask tracking and the best point-based tracking achieve comparable IoU scores, and (2) cases where segmentation mask tracking performs substantially better than point-based tracking.

These comparisons are illustrated in Fig. 1. As seen in these examples, when the surrounding tissues, such as connective tissue or fat, share similar appearance and texture with the gallbladder, the model often fails to maintain correct point tracking. The tracking points struggle to capture the deformable and indistinct

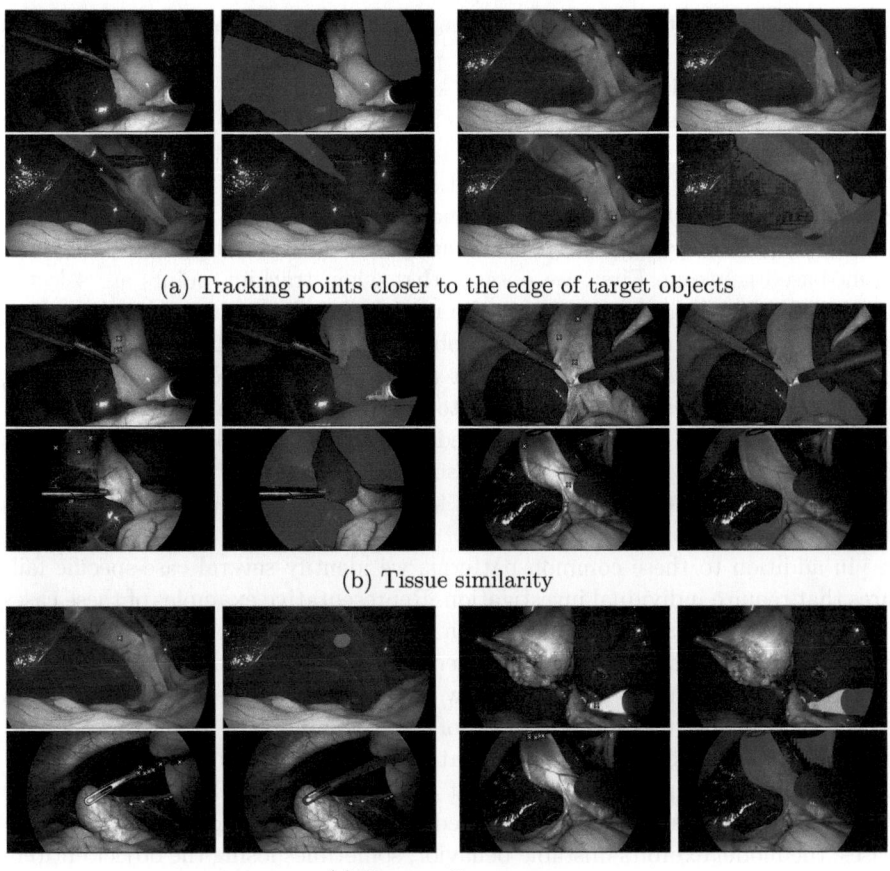

Fig. 2. Qualitative examples illustrating three distinct failure modes in point-based tracking for the gallbladder. For each image pair, the left image represents the input with tracking points and thee right one the segmentation output. (a) Failures due to tracking points placed near the object edges, causing the model to lose the target boundary. (b) Failures caused by tissue similarity, where surrounding structures confuse the model and lead to tracking drift. (c) Extraordinary cases that require case-specific investigation, such as partial object visibility or ambiguous visual cues. Incorrect predictions are highlighted in red and correctly predicted regions are highlighted using different colors for three objects (pink, blue, and cyan). (Color figure online)

boundaries of the organ, leading to the selection of adjacent regions into the segmentation.

On the other hand, in simpler scenes where the gallbladder has clear boundaries and is well-isolated from surrounding structures, point-based tracking performs competitively. In these cases, even a small number of points can successfully follow the object across frames, indicating that visual separation plays a critical role in tracking success.

Overall, these findings highlight a key failure mode of point-based tracking for anatomical structures: the model struggles when object boundaries are ambiguous or when object appearance blends with nearby tissues. Mask-based tracking, which has full spatial context, is less susceptible to these issues and remains consistently robust in such scenarios.

Upon qualitative inspection, we identify several recurring failure modes in point-based tracking. First, we observe that when tracking points are selected near the edges of the target object, the model often becomes confused, causing the tracked region to drift outside the object boundaries. This issue is observed not only for anatomical structures but also for surgical tools, where selecting points on tool edges sometimes leads to tracking leakage into the background or adjacent objects. Second, as discussed earlier, tissue similarity plays a major role in failure cases: when adjacent tissues closely resemble the tracked object in color or texture, the model struggles to distinguish between them, leading to tracking errors.

In addition to these common patterns, we identify several case-specific failures that require individual investigation. Representative examples of these cases are shown in Fig. 2. For instance, when tracking the L-hook electrocautery or the grasper, if the tracking point is placed only on the tip and not on the handle, or vice-versa, the model tends to follow just the tip (or the handle) due to its strong color contrast, ignoring the rest of the tool. Similarly, when tracking with a single point, we frequently observe that the model only tracks a small localized region, failing to capture the full extent of the object. Another common failure occurs when objects are tracked at the edge of the surgical field of view; in these cases, the model exhibits unstable behavior, sometimes losing the object entirely as it partially exits the frame.

3.1 Practical Recommendations for Point-Based Tracking

Based on our experimental findings, we provide the following practical suggestions for using video object segmentation models such as SAM2 in surgical videos when relying on tracking points:

1. **For anatomical structures:** Place several tracking points along the edges of the object rather than at the center. This reduces the likelihood of drift caused by boundary ambiguity and tissue similarity.
2. **For surgical instruments:** Place several points near the center of the instrument, as tools often appear very thin in the field of view. Additionally, distribute tracking points across visually distinct parts of the tool. For example, placing points on both the white/gray tip and the black handle of the grasper helps the model recognize that these parts belong to the same object.
3. **When initializing tracking near the edge of the field of view:** Exercise caution, as partial visibility and abrupt scene changes at the frame boundary can lead to tracking failures. Ensuring that points are placed on stable, clearly visible regions mitigates these risks.

4 Conclusion

In this work, we investigated the failure modes of point-based tracking in surgical video segmentation using SAM2. Through systematic experiments on cholecystectomy videos, we demonstrated that while point-based tracking can achieve competitive performance for surgical tools, it struggles with anatomical structures due to boundary ambiguity and tissue similarity. Our qualitative analysis further revealed that point placement plays a critical role in tracking success.

Future work should explore the use of *negative points*, which indicate areas that should not be tracked. These points could provide valuable context, helping the model avoid drift and improving tracking robustness in anatomically complex scenes. Due to the limited scope of this study, we did not investigate several important anatomical structures such as connective tissue and the liver, which frequently interact with surgical targets and contribute to visual ambiguity. Future research should expand the range of tracked objects to include these challenging tissues, enabling a more comprehensive understanding of model failure modes across diverse anatomical contexts.

References

1. Bansal, M., Kumar, M., Kumar, M., Kumar, K.: An efficient technique for object recognition using Shi-Tomasi corner detection algorithm. Soft. Comput. **25**(6), 4423–4432 (2021)
2. Chen, X.: Real-time semantic segmentation algorithms for enhanced augmented reality. J. Comput. Innov. **3**(1) (2023)
3. Dosovitskiy, A., et al.: An image is worth 16x16 words: transformers for image recognition at scale. arXiv preprint arXiv:2010.11929 (2020)
4. Gao, M., Zheng, F., Yu, J.J., Shan, C., Ding, G., Han, J.: Deep learning for video object segmentation: a review. Artif. Intell. Rev. **56**(1), 457–531 (2023)
5. Geetha, A.S., Hussain, M.: From SAM to SAM 2: exploring improvements in meta's segment anything model (2024). https://arxiv.org/abs/2408.06305
6. Hong, W.Y., Kao, C.L., Kuo, Y.H., Wang, J.R., Chang, W.L., Shih, C.S.: Cholecseg8k: a semantic segmentation dataset for laparoscopic cholecystectomy based on cholec80. arXiv preprint arXiv:2012.12453 (2020)
7. Jiaxing, Z., Hao, T.: SAM2 for image and video segmentation: a comprehensive survey (2025). https://arxiv.org/abs/2503.12781
8. Kaur, N.K., Kaur, U., Singh, D.: K-medoid clustering algorithm - a review. Int. J. Comput. Appl. Technol. **1**(1), 42–45 (2014)
9. Kitaguchi, D., Takeshita, N., Hasegawa, H., Ito, M.: Artificial intelligence-based computer vision in surgery: recent advances and future perspectives. Ann. Gastroenterol. Surg. **6**(1), 29–36 (2022). https://doi.org/10.1002/ags3.12513, https://onlinelibrary.wiley.com/doi/abs/10.1002/ags3.12513
10. Kulyabin, M., et al.: Segment anything in optical coherence tomography: SAM 2 for volumetric segmentation of retinal biomarkers. Bioengineering **11**(9), 940 (2024)
11. Liu, H., Zhang, E., Wu, J., Hong, M., Jin, Y.: Surgical SAM 2: real-time segment anything in surgical video by efficient frame pruning. arXiv preprint arXiv:2408.07931 (2024)

12. Loftus, T.J., et al.: Artificial intelligence and surgical decision-making. JAMA Surg. **155**(2), 148–158 (2020). https://doi.org/10.1001/jamasurg.2019.4917
13. Ma, J., et al.: MedSAM2: segment anything in 3D medical images and videos (2025). https://arxiv.org/abs/2504.03600
14. Mousavi, S.A., et al.: A reference-based approach for tumor size estimation in monocular laparoscopic videos. In: Wu, J., Qin, W., Li, C., Kim, B. (eds.) CMMCA 2024. LNCS, vol. 15181, pp. 11–20. Springer, Cham (2024). https://doi.org/10.1007/978-3-031-73360-4_2
15. Obuchowicz, R., Strzelecki, M., Piórkowski, A.: Clinical applications of artificial intelligence in medical imaging and image processing—a review. Cancers **16**(10) (2024). https://doi.org/10.3390/cancers16101870, https://www.mdpi.com/2072-6694/16/10/1870
16. Ozbulak, U., et al.: Revisiting the evaluation bias introduced by frame sampling strategies in surgical video segmentation using SAM2 (2025). https://arxiv.org/abs/2502.20934
17. Rajič, F., Ke, L., Tai, Y.W., Tang, C.K., Danelljan, M., Yu, F.: Segment anything meets point tracking. In: 2025 IEEE/CVF Winter Conference on Applications of Computer Vision (WACV), pp. 9302–9311. IEEE (2025)
18. Ravi, N., et al.: SAM 2: segment anything in images and videos. arXiv preprint arXiv:2408.00714 (2024)
19. Vaswani, A., et al.: Attention is all you need. In: Advances in Neural Information Processing Systems, vol. 30 (2017)

Deep Biomechanically-Guided Interpolation for Keypoint-Based Brain Shift Registration

Tiago Assis[1](✉), Ines P. Machado[2,3], Benjamin Zwick[4], Nuno C. Garcia[1], and Reuben Dorent[5,6]

[1] LASIGE, Faculdade de Ciências da Universidade de Lisboa,
1749-016 Lisbon, Portugal
tassis@lasige.di.fc.ul.pt
[2] CRUK Cambridge Centre, University of Cambridge, Cambridge, UK
[3] Department of Oncology, University of Cambridge, Cambridge, UK
[4] Intelligent System for Medicine Laboratory (ISML), School of Mechanical Engineering, The University of Western Australia, Perth, WA 6009, Australia
[5] MIND Team, Inria Saclay, Université Paris-Saclay, Palaiseau, France
reuben.dorent@inria.fr
[6] Sorbonne Université, Institut du Cerveau - Paris Brain Institute - ICM, CNRS, Inria, Inserm, AP-HP, Hôpital de la Pitiè Salpêtrière, 75013 Paris, France

Abstract. Accurate compensation of brain shift is critical for maintaining the reliability of neuronavigation during neurosurgery. While keypoint-based registration methods offer robustness to large deformations and topological changes, they typically rely on simple geometric interpolators that ignore tissue biomechanics to create dense displacement fields. In this work, we propose a novel deep learning framework that estimates dense, physically plausible brain deformations from sparse matched keypoints. We first generate a large dataset of synthetic brain deformations using biomechanical simulations. Then, a residual 3D U-Net is trained to refine standard interpolation estimates into biomechanically guided deformations. Experiments on a large set of simulated displacement fields demonstrate that our method significantly outperforms classical interpolators, reducing by half the mean square error while introducing negligible computational overhead at inference time. Code available at: https://github.com/tiago-assis/Deep-Biomechanical-Interpolator.

Keywords: Displacement Interpolation · Biomechanical Modeling · Image Registration · Physically-Guided Deep Learning

1 Introduction

Image registration is a fundamental task in image-guided surgery, enabling the spatial alignment of preoperative data with intraoperative anatomy. In neurosurgical procedures, accurate registration is particularly critical, as it supports neuronavigation systems that guide the surgeon based on preoperative Magnetic Resonance Imaging (MRI). However, the reliability of these systems degrades as surgery progresses due to intraoperative brain deformations, often referred to as *brain shift*, caused by gravity, tissue resection, and cerebrospinal fluid loss [9].

To compensate for brain shift, a wide range of registration methods leveraging intraoperative imaging modalities such as intraoperative MRI (iMRI) and ultrasound (iUS) have been proposed, including learning-based [3,6,18,19] and non-learning-based [1,20,24] methods. These techniques typically align pre- and intraoperative images by optimizing intensity-based similarity metrics. However, they often struggle in challenging registration scenarios involving (1) large intensity distribution gaps between pre- and intraoperative modalities (e.g., MRI to iUS), (2) large deformations, and (3) topological changes due to tissue resection.

Keypoint-based registration methods have recently gained traction as a competitive alternative [10,21,22,25]. By relying on sparse correspondences rather than voxel-wise similarity, these methods are more robust to large deformations, partial fields of view, and topological changes. They also offer interpretable outputs, as matched keypoints can be directly visualized and assessed. However, keypoint-based methods typically rely on simple geometric interpolators, such as thin-plate splines or linear models, to propagate sparse displacements into dense displacement fields. These interpolators ignore the biomechanical properties of brain tissue, which can result in physically unrealistic deformations.

In this work, we propose a novel deep learning framework for estimating dense and physically plausible brain deformations from sparse matched keypoints between pre- and intra-operative images (Fig. 1). First, we construct a large-scale dataset of synthetic brain surgical deformations using biomechanical simulations. Second, we simulate matched keypoints by extracting keypoints using 3D SIFT and pairing them with ground-truth displacements from the synthetic deformations. Third, we develop a deep, biomechanically guided interpolator based on a residual 3D U-Net that refines standard interpolation estimates using the preoperative data. Finally, extensive experiments were conducted on simulated brain deformations, outperforming standard interpolation methods significantly in terms of displacement error, with negligible computational overhead.

2 Methods

2.1 Overview and Problem Setting

In this work, we assume access to a preoperative MRI $I_{\text{pre}} \in \mathbb{R}^{D \times W \times H}$, where D denotes the depth, W the width, H the height, and a sparse set of M matched keypoints $\{(\boldsymbol{x}_i, \boldsymbol{y}_i)\}_{i=1}^{M}$, where $(\boldsymbol{x}_i, \boldsymbol{y}_i) \in \mathbb{R}^3 \times \mathbb{R}^3$ represent corresponding 3D anatomical locations in the preoperative and intraoperative spaces, respectively. The matched keypoints may be obtained either manually or automatically from any intraoperative imaging modality, such as iMRI or ultrasound. At each keypoint $\boldsymbol{x}_i$, the displacement vector $\boldsymbol{d}_i = \boldsymbol{y}_i - \boldsymbol{x}_i$ captures the local brain displacement occurring during surgery. Our objective is to estimate a *dense* and

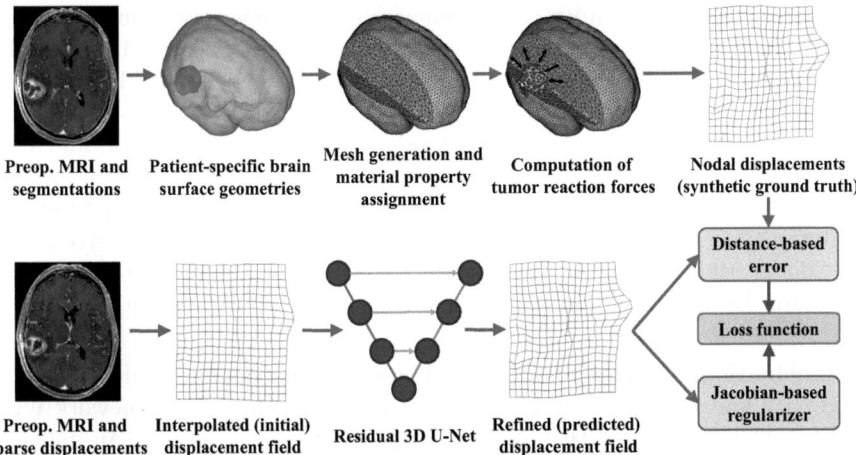

Fig. 1. Overview of the proposed framework. A biomechanical simulation generates synthetic ground truth displacement fields using preoperative MRI and joint segmentation of tumor and surrounding structures. Sparse intraoperative keypoint displacements are interpolated to form an initial estimate, which is refined by a residual 3D U-Net. The final displacement field is supervised using voxel-wise error and a Jacobian-based regularization loss.

physically plausible displacement field $\phi \in \mathbb{R}^{3 \times D \times W \times H}$ that estimates the surgical brain deformations. To this end, we propose to train a deep interpolator f_θ, parameterized by learnable parameters θ, using biomechanical simulations.

2.2 Ground-Truth Brain Deformation Dataset

The goal of the deep interpolator f_θ is to estimate a dense displacement field from a sparse set of M displacement vectors $\{d_i\}_{i=1}^{M}$. In the absence of ground-truth dense deformations in clinical data, we propose to train our interpolator using synthetic deformations generated through biomechanical simulations. Specifically, we leverage the biomechanical framework introduced in [26], which simulates brain deformations induced by tumor resection using the Meshless Total Lagrangian Explicit Dynamics (MTLED) algorithm [14]. This approach employs a Total Lagrangian formulation with explicit time integration to realistically model intraoperative tissue deformation.

Brain Tumor Dataset. We used the UPENN-GBM dataset [2], which comprises preoperative multi-parametric brain MRIs and tumor segmentations from $N = 162$ patients diagnosed with de novo glioblastoma. For this study, we employed the contrast-enhanced T1 (ceT$_1$) scans as the preoperative MRI I_{pre}, along with their manual segmentations of the tumor core.

Patient-Specific Geometry. The biomechanical framework requires patient-specific brain geometry, including the surfaces of the tumor core and surrounding structures: brain parenchyma, cerebrospinal fluid (CSF), and skull. While the segmentation of brain regions in the presence of tumors can be obtained using dedicated frameworks [7,11], these methods do not delineate the CSF and skull. Instead, we use SynthSeg [4], a tool not specifically designed for pathological data, to segment the parenchyma and CSF on ceT$_1$ images. Then, these segmentations are merged with the tumor core segmentation and converted using the "Model Maker" module in 3D Slicer [8] into triangulated surface models, which are the main input to the biomechanical simulation pipeline.

Biomechanical Simulations. Surgical brain deformations are primarily driven by gravity and tissue resection. Following the biomechanical framework introduced in [26], we assume that prior to craniotomy, the brain is in an unloaded state in which gravitational forces are balanced by the buoyancy of intracranial fluids. Craniotomy disrupts this equilibrium due to pressure release and CSF drainage, resulting in gravity-induced brain deformation.

The meshless approach uses a cloud of points for spatial discretization and tetrahedral integration cells, simplifying grid construction compared to traditional finite element methods. Two biomechanical brain models are constructed: a pre-resection model that includes the tumor, and a post-resection model in which tumor nodes and their connectivity are removed to simulate the resection cavity. We follow the material parameters for the Ogden model used for Patient 1 in [26], which are assigned using fuzzy tissue classification, allowing probabilistic tissue labeling without explicit segmentations. The parenchyma is modeled as nearly incompressible, the tumor stiffer with a shear modulus three times that of healthy tissue, and CSF as highly compressible to reflect fluid drainage dynamics.

Then, gravity-induced deformation is simulated by applying the resulting unbalanced forces to the pre-resection model. To model post-resection deformation, internal reaction forces at the tumor-parenchyma interface are computed and applied in the opposite direction to the post-resection model.

Unlike the original framework, where the gravity vector was manually defined, we propose an automated estimation method. Specifically, we derive the base gravity direction from the surface point nearest to the tumor center, based on the assumption that surgeons typically select the shortest access path to the tumor. To account for variability in patient positioning and surgical approach, we generate K plausible gravity vectors by perturbing the base direction by up to $\pm 10°$ along each spatial axis. This leads to the dataset $\mathcal{D}_{\text{total}} = \{(I_{\text{pre}}^{(j)}, \phi_{\text{gt}}^{(k,j)})_{k=1}^{K}\}_{j=1}^{N}$, containing K distinct displacement fields for each of the N preoperative MRI.

2.3 Synthetic Matched Keypoints Strategy

To simulate the acquisition of sparse sets of M displacement vectors $\{\boldsymbol{d}_i\}_{i=1}^{M}$ for a preoperative MRI I_{pre}, we adopt the following strategy: (1) we extract keypoints from I_{pre} using the widely used 3D SIFT algorithm [5], and (2) retrieve their associated displacement vectors using the ground-truth synthetic displacement field ϕ_{gt}. The 3D SIFT algorithm automatically identifies anatomically meaningful landmarks that are robust to variations in intensity and structure. Since SIFT typically produces hundreds to thousands of keypoints, we randomly sample M keypoints uniformly from the detected set.

2.4 Deep Physically-Inspired Interpolator

To design our deep physically-inspired interpolator, we use a denoising approach. Given a sparse set of displacement vectors $\{d_i\}_{i=1}^{M}$, we first compute an initial dense displacement field $\phi_{init} \in \mathbb{R}^{3 \times D \times W \times H}$ using a standard interpolation technique such as linear (L) or thin-plate spline (TPS) interpolation. This initial estimate ϕ_{init} is then refined by a deep interpolator $f_\theta : \left(\mathbb{R}^{D \times W \times H}, \mathbb{R}^{3 \times D \times W \times H} \right) \mapsto \mathbb{R}^{3 \times D \times W \times H}$ conditioned on the preoperative image I_{pre} to approximate the ground-truth displacement field ϕ_{gt}, i.e. $f_\theta(I_{pre}, \phi_{init}) \approx \phi_{gt}$.

Training Procedure. The deep interpolator f_θ is trained under full supervision using the synthetic ground-truth displacement fields. At each training iteration, we randomly sample a training preoperative image I_{pre} with its 3D SIFT keypoints and a pre-computed ground-truth displacement field ϕ_{gt}. Then, we sample a set of M sparse displacements $\{d_i\}_{i=1}^{M}$ and compute on-the-fly an initial dense displacement field ϕ_{init} using a standard interpolation technique (L or TPS). The network f_θ is trained to minimize the mean squared error (MSE) between the predicted $\phi_{pred} = f_\theta(I_{pre}, \phi_{init})$ and true displacement fields ϕ_{gt} over the image domain. To encourage smooth displacements in non-resected brain regions, we introduce an additional Jacobian determinant regularization that encourages a local orientation consistency constraint on the estimated displacement field. The total loss function $\mathcal{L}$ then becomes:

$$\mathcal{L} \triangleq \|\phi_{pred} - \phi_{gt}\|_2^2 + \frac{\lambda_{reg}}{|\Omega_{healthy}|} \sum_{x \in \Omega_{healthy}} \text{ReLU}\left(-\det J_{I+\phi_{pred}}(x)\right), \quad (1)$$

where λ_{reg} weights the regularization term, $\Omega_{healthy}$ denotes the non-tumorous brain area and $J_{I+\phi_{pred}}$ is the Jacobian matrix of the deformation field. ReLU ensures that only negative Jacobian determinants are penalized.

Residual Network Architecture. Motivated by denoising diffusion models [12], which have shown that learning to predict the noise in a noisy signal leads to better performance than predicting the clean signal directly, our network predicts a residual displacement ϵ_θ, such that $f_\theta(I_{pre}, \phi_{init}) = \phi_{init} + \epsilon_\theta(I_{pre}, \phi_{init})$. The residual network f_θ is a 3D U-Net architecture variant of [17]. At each resolution level, residual blocks are employed with spatial and channel-wise squeeze-and-excitation (SE) [23] modules. Downsampling in the encoder path is performed using max-pooling, while upsampling in the decoder path is achieved via transposed convolutions. Same-size feature maps from the encoder are merged

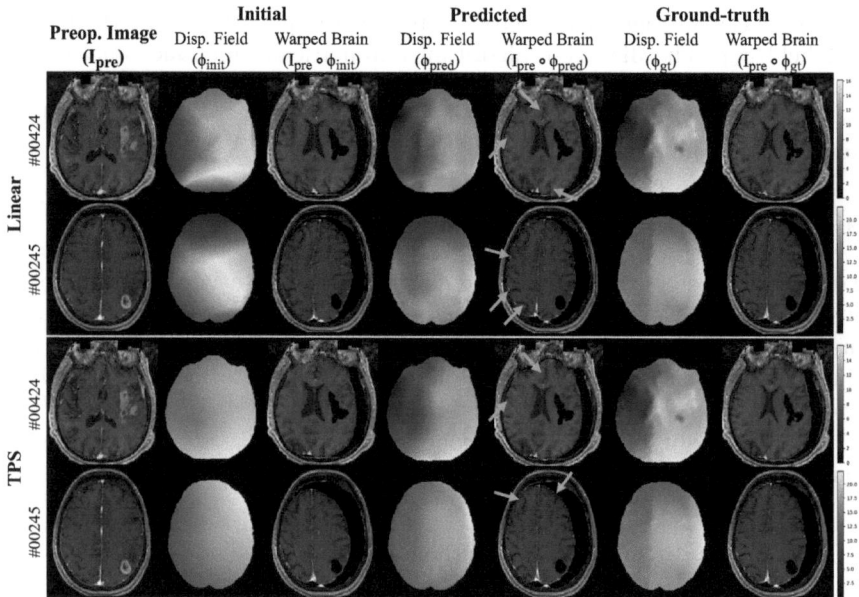

Fig. 2. Qualitative comparison of displacement fields and resulting warped brain anatomies. The displacement fields are colored by vector magnitudes, and green arrows highlight the most noticeable improvements over the linear and TPS baselines. (Color figure online)

with decoder features through element-wise summation rather than concatenation. The network comprises 4 resolution levels, starting with 32 feature channels and doubling at each downsampling stage up to a maximum of 256 channels, while the spatial resolution is halved. Each convolution within the residual blocks is followed by instance normalization and a LeakyReLU activation function with a negative slope of 10^{-2}, except for the final convolution in each block, where activation is applied after the residual summation. The SE modules are then applied at the end of each block, resulting in a network with 7.3M parameters.

3 Experiments

Data. We evaluated our method using the UPENN-GBM dataset, where 204 synthetic ground-truth displacement fields were generated via biomechanical simulations for 162 unique patients ($K = 1$-3 simulations per case). The dataset was split into 121 training, 16 validation, and 25 test cases, following a 75:10:15 ratio. For each case, a ground-truth displacement field was randomly selected from the set of simulations with different gravity-induced brain shifts, and a random set of keypoints was sampled to initialize the displacement field.

Implementation. All inputs were cropped to a fixed size of $160 \times 192 \times 144$. Preoperative ce$T_1$ images were normalized by subtracting the mean and dividing

Table 1. Quantitative evaluation of different approaches using linear (L) and thin-plate spline (TPS) interpolation using $M = 20$ keypoints. R denotes the residual architecture and J the Jacobian regularization term. Mean and standard deviation are reported. Statistical significance was determined using a Bonferroni-corrected paired Wilcoxon signed-rank test, * indicating statistically significant improvements ($p < 0.01$) for Ours.

Method	MSE (mm^2) ↓		Max Error ↓	HD95 ↓	%$\|J_\phi\| < 0$ ↓	Time ↓
	Brain	Edema	(mm)	(mm)	(%)	(s)
L (baseline)	10.7* (5.4)	7.6 (5.6)	28.3 (8.0)	3.7* (1.1)	0.74* (0.11)	1.81 (0.02)
Ours (L)$_{w/o\ R+J}$	4.2 (2.2)	10.9* (6.6)	27.8 (8.5)	2.8 (0.7)	0.97* (0.20)	-
Ours (L)$_{w/o\ J}$	3.7 (1.7)	9.4 (7.1)	27.3 (8.4)	2.8 (0.5)	1.16* (0.25)	-
Ours (L)	3.7 (1.6)	6.4 (3.0)	26.2 (8.2)	2.7 (0.6)	0.64 (0.21)	1.81 (0.02)
TPS (baseline)	6.5* (2.6)	6.4 (5.5)	25.4* (6.9)	2.8 (0.7)	0.64* (0.21)	0.58 (0.01)
Ours (TPS)$_{w/o\ R+J}$	4.6* (2.1)	10.6* (7.5)	26.4 (8.7)	2.9 (1.0)	1.47* (0.36)	-
Ours (TPS)$_{w/o\ J}$	3.5 (1.6)	7.2 (5.2)	23.4 (5.2)	3.3 (0.7)	0.99* (0.14)	-
Ours (TPS)	3.4 (1.6)	5.9 (3.3)	22.7 (4.9)	3.1 (0.5)	0.59 (0.22)	0.59 (0.01)

by the standard deviation. Data augmentation was applied only to images during training and included Gaussian noise and blur, intensity adjustments (brightness, contrast, gamma), and simulated low resolution. These augmentations followed the strategies used in nnU-Net [13] to improve generalization capability. Our approach was trained using the Adam optimizer with a learning rate of 5×10^{-4}, a batch size of 1, and for 100 epochs. To ensure anatomical relevance for all methods, interpolated displacements in the background or skull are set to zero. A value of $\lambda_{\text{reg}} = 50$ was chosen by performing a grid search on the validation set, which provided the best balance between performance and regularization.

Metrics. To assess the performance of our method, we evaluated the predicted displacement fields using several complementary metrics. The mean squared error (**MSE**) in mm^2 was computed between the predicted and ground-truth displacement fields, both within the whole brain and specifically within the edematous tumor region, to quantify the overall accuracy and the accuracy near the resected area, respectively. We additionally reported the maximum Euclidean error (**Max Error**) in mm, capturing the worst-case deviation in the predicted displacements. To assess the geometric alignment of brain structures, we computed the 95th percentile Hausdorff distance (**HD95**) between the brain segmentations warped by the predicted and ground-truth displacement fields. To evaluate the anatomical plausibility of the deformations, the percentage of voxels with non-positive Jacobian determinant values (**%$|J_\phi| < 0$**) was computed. Finally, we reported the inference time (**Time**) to evaluate the computational efficiency of our method in comparison to the baselines.

Baseline Interpolation Methods and Ablation Study. Two widely-used interpolation methods were used in our experiments for comparison and initialization: (1) linear interpolation (L) via a 3D Delaunay triangulation approach using a publicly available differentiable implementation [15] and (2) a

Table 2. Impact of the number of M matched keypoints in terms of MSE (mm^2).

Method	$M=5$	$M=10$	$M=15$	$M=20$	$M=50$
L (baseline)	17.2 (9.1)	13.3 (6.5)	11.4 (5.9)	10.7 (5.4)	6.4 (3.1)
Ours (L)	**7.8 (4.3)**	**4.8 (2.5)**	**4.7 (2.7)**	**4.0 (2.0)**	**2.3 (1.1)**
TPS (baseline)	18.0 (12.4)	10.2 (4.8)	7.2 (2.9)	6.5 (2.6)	3.8 (1.6)
Ours (TPS)	**11.0 (8.4)**	**5.7 (3.3)**	**4.4 (2.5)**	**3.4 (1.6)**	**2.1 (1.1)**

thin-plate spline (TPS) approach using a public implementation [25]. A regularization weight λ_{tps} of 0.1 was used for TPS, which achieved the best empirical results on the validation set. Both interpolation methods serve as non-learning-based baselines that do not incorporate anatomical context or physical priors.

We also performed an ablation study to evaluate the impact of three key components of our method: (1) the choice of interpolator for displacement field initialization, (2) a residual network architecture (R), and (3) the Jacobian regularization term (J). All configurations used a fixed set of 20 keypoints per case.

Results. Results are shown in Table 1 and Fig. 2. Our method consistently outperformed baseline interpolators, reducing whole-brain MSE by up to 47% (3.07 mm^2) with TPS and 65% (6.96 mm^2) with linear interpolation. Improvements in the edema were more modest but with reduced variability, suggesting a more stable approach. The residual architecture improved accuracy across the board, especially in the edema region, by enabling the model to learn finer corrections. Notably, omitting residual learning with TPS initialization led to increased voxel folding, which can be attributed to TPS's lack of flexibility to adapt to fine-grained deformations, leading the network to overcompensate. Adding a Jacobian regularizer improved deformation smoothness, reducing non-invertible mappings by 40 to 45% (+0.40 to +0.52pp), without compromising accuracy. Inference time increased negligibly compared to baseline methods (+10ms).

Impact of Number of Keypoints. Finally, we analyzed the impact of the number of input keypoints M, varying it from 5 to 50 (Table 2). As expected, increasing the number of keypoints led to lower MSE, as the interpolation benefits from more accurate and localized displacement observations. At low keypoint counts (e.g., 5), TPS interpolation performed poorly, likely due to instability with limited control points. In contrast, linear interpolation demonstrated greater robustness in such settings. However, with a higher number of keypoints (e.g., 50), TPS produced smoother and more accurate results, outperforming linear interpolation. This illustrates a trade-off between robustness and smoothness that depends on the spatial density of keypoints. Notably, in all cases, our deep interpolator significantly improved upon the initial interpolation, reducing error regardless of the number of points M and the interpolation method.

4 Conclusion

We introduced a deep learning framework for estimating dense and physically plausible brain deformations from sparse keypoint correspondences between pre- and intra-operative images. To enable supervised training, we built a large dataset of synthetic brain deformations using biomechanical simulations across a large number of cases. Sparse keypoints were simulated using 3D SIFT and paired with ground-truth displacements to mimic intraoperative correspondences. A residual 3D U-Net was trained to refine standard interpolation fields into biomechanically-guided deformations, guided by the preoperative image and regularized by a Jacobian-based constraint. Our experiments demonstrate that the proposed method consistently outperforms classical interpolators in accuracy, without incurring significant computational cost at inference. Future work will explore the generalization capabilities of our framework on additional datasets, such as the ReMIND dataset [16]. We also aim to extend the method's robustness to handle imperfect correspondences and varying numbers of matched keypoints. Finally, we plan to apply our approach to image registration for surgical guidance.

Acknowledgments. Tiago Assis received a grant from Fundação para a Ciência e a Tecnologia (FCT), I.P./MCTES, through national funds (PIDDAC) under the Strategic Funding 2020–23 programme (Grant UIDB/00408/2020). Reuben Dorent received a Marie Skłodowska-Curie grant, No. 101154248 (project: SafeREG). This activity has been supported by the Western Australian Future Health Research and Innovation Fund (Grant ID WANMA/Ideas2023-24/13).

Disclosure of Interests. The authors have no competing interests to declare that are relevant to the content of this article.

References

1. Avants, B.B., Epstein, C.L., Grossman, M., Gee, J.C.: Symmetric diffeomorphic image registration with cross-correlation: evaluating automated labeling of elderly and neurodegenerative brain. Med. Image Anal. **12**(1), 26–41 (2008)
2. Bakas, S., et al.: Multi-parametric magnetic resonance imaging (mpMRI) scans for de novo Glioblastoma (GBM) patients from the University of Pennsylvania Health System (UPENN-GBM). The Cancer Imaging Archive (2021)
3. Balakrishnan, G., Zhao, A., Sabuncu, M.R., Guttag, J., Dalca, A.V.: Voxelmorph: a learning framework for deformable medical image registration. IEEE Trans. Med. Imaging **38**(8), 1788–1800 (2019)
4. Billot, B., et al.: SynthSeg: segmentation of brain MRI scans of any contrast and resolution without retraining. Med. Image Anal. **86**, 102789 (2023)
5. Chauvin, L., et al.: Neuroimage signature from salient keypoints is highly specific to individuals and shared by close relatives. Neuroimage **204**, 116208 (2020)
6. de Vos, B.D., Berendsen, F.F., Viergever, M.A., Staring, M., Išgum, I.: End-to-end unsupervised deformable image registration with a convolutional neural network. In: Cardoso, M.J., et al. (eds.) DLMIA/ML-CDS -2017. LNCS, vol. 10553, pp. 204–212. Springer, Cham (2017). https://doi.org/10.1007/978-3-319-67558-9_24

7. Dorent, R., et al.: Learning joint segmentation of tissues and brain lesions from task-specific hetero-modal domain-shifted datasets. Med. Image Anal. **67**, 101862 (2021)
8. Fedorov, A., et al.: 3D Slicer as an image computing platform for the Quantitative Imaging Network. Magn. Reson. Imaging **30**(9), 1323–1341 (2012)
9. Gerard, I.J., Kersten-Oertel, M., Petrecca, K., Sirhan, D., Hall, J.A., Collins, D.L.: Brain shift in neuronavigation of brain tumors: a review. Med. Image Anal. **35**, 403–420 (2017)
10. Heinrich, M.P., Hansen, L.: Voxelmorph++ going beyond the cranial vault with keypoint supervision and multi-channel instance optimisation. In: Hering, A., Schnabel, J., Zhang, M., Ferrante, E., Heinrich, M., Rueckert, D. (eds.) WBIR 2022. LNCS, vol. 13386, pp. 85–95. Springer, Cham (2022). https://doi.org/10.1007/978-3-031-11203-4_10
11. Himmetoglu, M., Ciernik, I., Konukoglu, E.: Learning to segment anatomy and lesions from disparately labeled sources in brain MRI. arXiv preprint arXiv:2503.18840 (2025)
12. Ho, J., Jain, A., Abbeel, P.: Denoising diffusion probabilistic models. In: Advances in Neural Information Processing Systems, vol. 33, pp. 6840–6851 (2020)
13. Isensee, F., Jaeger, P.F., Kohl, S.A., Petersen, J., Maier-Hein, K.H.: nnU-Net: a self-configuring method for deep learning-based biomedical image segmentation. Nat. Methods **18**(2), 203–211 (2021)
14. Joldes, G., et al.: Suite of meshless algorithms for accurate computation of soft tissue deformation for surgical simulation. Med. Image Anal. **56**, 152–171 (2019)
15. Joutard, S., Dorent, R., Ourselin, S., Vercauteren, T., Modat, M.: Driving points prediction for abdominal probabilistic registration. In: Lian, C., Cao, X., Rekik, I., Xu, X., Cui, Z. (eds.) MLMI 2022. LNCS, vol. 13583, pp. 288–297. Springer, Cham (2022). https://doi.org/10.1007/978-3-031-21014-3_30
16. Juvekar, P., et al.: Remind: the brain resection multimodal imaging database. Sci. Data **11**(1), 494 (2024)
17. Lee, K., Zung, J., Li, P., Jain, V., Seung, H.S.: Superhuman accuracy on the SNEMI3D connectomics challenge. arXiv preprint arXiv:1706.00120 (2017)
18. Mok, T.C.W., Chung, A.C.S.: Large deformation diffeomorphic image registration with Laplacian pyramid networks. In: Martel, A.L., et al. (eds.) MICCAI 2020. LNCS, vol. 12263, pp. 211–221. Springer, Cham (2020). https://doi.org/10.1007/978-3-030-59716-0_21
19. Mok, T.C., Chung, A.C.: Unsupervised deformable image registration with absent correspondences in pre-operative and post-recurrence brain tumor MRI scans. In: Wang, L., Dou, Q., Fletcher, P.T., Speidel, S., Li, S. (eds.) MICCAI 2022. LNCS, vol. 13436, pp. 25–35. Springer, Cham (2022). https://doi.org/10.1007/978-3-031-16446-0_3
20. Ou, Y., Sotiras, A., Paragios, N., Davatzikos, C.: DRAMMS: deformable registration via attribute matching and mutual-saliency weighting. Med. Image Anal. **15**(4), 622–639 (2011)
21. Rasheed, H., et al.: Learning to match 2D keypoints across preoperative mr and intraoperative ultrasound. In: Gomez, A., Khanal, B., King, A., Namburete, A. (eds.) ASMUS 2024. LNCS, vol. 15186, pp. 78–87. Springer, Cham (2024). https://doi.org/10.1007/978-3-031-73647-6_8
22. Rister, B., Horowitz, M.A., Rubin, D.L.: Volumetric image registration from invariant keypoints. IEEE Trans. Image Process. **26**(10), 4900–4910 (2017)

23. Roy, A.G., Navab, N., Wachinger, C.: Concurrent spatial and channel 'squeeze & excitation' in fully convolutional networks. In: Frangi, A.F., Schnabel, J.A., Davatzikos, C., Alberola-López, C., Fichtinger, G. (eds.) MICCAI 2018. LNCS, vol. 11070, pp. 421–429. Springer, Cham (2018). https://doi.org/10.1007/978-3-030-00928-1_48
24. Vercauteren, T., Pennec, X., Perchant, A., Ayache, N.: Diffeomorphic demons: efficient non-parametric image registration. Neuroimage **45**(1), S61–S72 (2009)
25. Wang, A.Q., Evan, M.Y., Dalca, A.V., Sabuncu, M.R.: A robust and interpretable deep learning framework for multi-modal registration via keypoints. Med. Image Anal. **90**, 102962 (2023)
26. Yu, Y., et al.: Automatic framework for patient-specific modelling of tumour resection-induced brain shift. Comput. Biol. Med. **143**, 105271 (2022)

Nested ResNet: A Vision-Based Method for Detecting the Sensing Area of a Drop-In Gamma Probe

Songyu Xu[1], Yicheng Hu[1], Jionglong Su[2], Daniel S. Elson[1], and Baoru Huang[1,3(✉)]

[1] The Hamlyn Center for Robotic Surgery, Imperial College London, London, UK
[2] School of AI and Advanced Computing, Xia'an Jiaotong-Liverpool University, Suzhou, China
[3] Liverpool University, Liverpool, UK
baoru.huang@liverpool.ac.uk

Abstract. Drop-in gamma probes are widely used in robotic-assisted minimally invasive surgery (RAMIS) for lymph node detection. However, these devices only provide audio feedback on signal intensity and lack the visual feedback necessary for precise localization. Previous work attempted to predict the sensing area location using laparoscopic images, but the prediction accuracy was unsatisfactory. We propose a three-branch deep learning framework that leverages stereo laparoscopic images as input to a Nested ResNet main branch, with depth estimation (via transfer learning) and orientation guidance (through probe axis sampling) as additional guidance. Our approach has been evaluated on a publicly available dataset, demonstrating superior performance over previous methods. Quantitatively, our method yields a 22.10% decrease in 2D mean error and a 41.67% reduction in 3D mean error. Qualitative analyses further confirm the enhanced precision of our solution. This advancement enables real-time visual feedback during gamma probe use in RAMIS, improving surgical localization accuracy and reliability. Our code is available at: https://github.com/Songyu-Xu/Nested-ResNet.

Keywords: Image-guided surgery · Minimally invasive surgery · Drop-in gamma probe · Deep learning

1 Introduction

Cancer remains a leading cause of death globally [20], although early diagnosis and treatment can improve survival rates. Robotic-assisted minimally invasive surgery (RAMIS) plays a vital role in oncological treatment, where accurate localization of tumors is critical. In laparoscopic radio-guided procedures, patients are pre-injected with radiopharmaceutical agents that can be designed to selectively accumulate in cancerous tissues or lymph nodes [16]. Gamma probes or cameras are then used to detect these radioactive signals, enabling effective intraoperative localization [5].

The SENSEI® (Lightpoint Medical Ltd) [12] is a recently developed drop-in gamma probe that is designed to be compatible with both standard non articulated (e.g., laparoscopic Johan grasper) and articulated (e.g., da Vinci Pro-Grasp) tissue graspers [9] that can be manipulated by surgical robots [3]. The gamma probe can produce auditory feedback according to the signal intensity detected through a small window at its tip. As the intensity increases, the console will produce faster beeps, therefore aiding surgeons in localizing cancerous tissue. However, these lack spatial cues and requires the surgeon to recall areas of strong signal, which can lead to discrepancies between perceived and actual signal locations. This may result in repeated scanning, increasing workload and procedure duration. Therefore, a visual feedback system is needed to indicate the probe's sensing area even when it is not in contact with tissue.

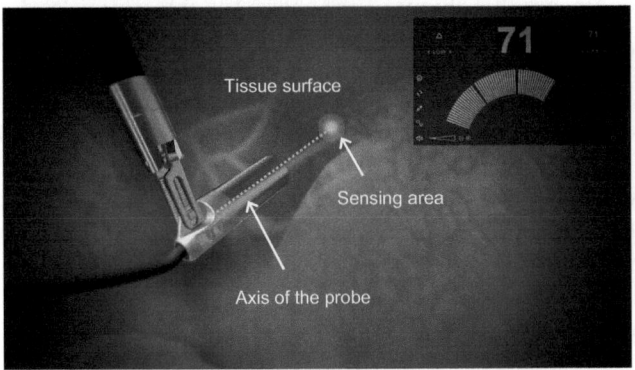

Fig. 1. The SENSEI® gamma probe working scenario and the definition of the sensing area.

To address this, a visual feedback system is needed to estimate the probe sensing area, defined geometrically as the intersection between the probe axis and the tissue surface (Fig. 1) [8]. Although 3D reconstruction offers a solution, it requires accurate geometry and pose data, which is difficult to obtain intraoperatively. A 2D image-based method proposed by Huang et al. [8] simplified the problem via regression, but showed limited accuracy.

We propose a novel deep learning framework based on a feature-guided Nested ResNet to predict the gamma probe's sensing area from laparoscopic images. Through extensive experiments, our method achieves superior accuracy over previous approaches. This prediction serves as a foundational step in enabling downstream tasks such as tracking high-activity regions, which can further reduce repeat scanning and improve intraoperative efficiency.

2 Methodology

The proposed framework, shown in Fig. 2, takes laparoscopic images as input, supplemented by depth maps and probe axis points. A Nested ResNet extracts features from the images, while a CNN and an MLP process the depth maps and probe axis data, respectively. Features from all three branches are concatenated and fed into a four-layer MLP with decreasing neuron counts (512, 256, 64, 2) to generate the final prediction. The following sections describe each branch in detail.

2.1 Depth Estimation

We utilized a disparity estimation model [21] that can predict the disparity map from a pair of stereo images. This method accepts stereo image pairs as the input and estimates the disparity map using a deep learning approach. It employs convolutional neural networks (CNN) to extract features from each image independently and utilizes a transformer-based model to learn feature matching on corresponding images [21]. Figure 2 (a) shows the integration of depth maps in our model. We utilize the disparity estimation model that has been pretrained on three stereo datasets for predicting the disparity map: KITTI Stereo [14]; Middlebury [18]; and ETH3D Stereo [19]. After obtaining the disparity D, the per-pixel depth Z can be calculated using $Z = fb/D$, where f is the camera's focal length and b is the baseline. The depth map is processed by a four-layer CNN, followed by a linear layer that reduces the feature size to 128. A ReLU [15] activation function is then applied. The extracted features are subsequently concatenated with those from the other branches.

2.2 Orientation Guidance Using Central Axis

As illustrated in Fig. 1, the sensing area is the intersection of the probe axis and the tissue surface. The probe axis contains key information about the orientation of the sensing area, hence we address the orientation information by introducing the extracted probe axis as a supplementary input to our network (Fig. 2 (c)). Firstly, the mask of the probe is generated using a SAM-based [11] software on the Coffbea dataset [8] to ensure a reliable ground truth mask, and the DeepLabv3 model [1] is used for automatic segmentation. Subsequently, Principal Component Analysis (PCA) [13] can determine the central axis of the segmentation mask as the greatest variance occurs near the axis of symmetry [8]. A random sample of $n = 50$ points along this axis, denoted as $(u_1, v_1), (u_2, v_2), \ldots, (u_n, v_n)$, is used to represent the probe orientation, and a two-layer MLP, with 256 and 128 neurons, is applied to extract axis features. These are then concatenated with features from the other two branches.

2.3 Nested ResNet

While the depth branch (Fig. 2 (a)) and the axis branch (Fig. 2 (c)) provide key features for understanding the spatial relationships, it is necessary to extract features directly from the original RGB images as they contain all information (Fig. 2 (b)). One challenge of the intersection estimation task is that the sensing area does not have any pattern characteristics but is jointly defined by the

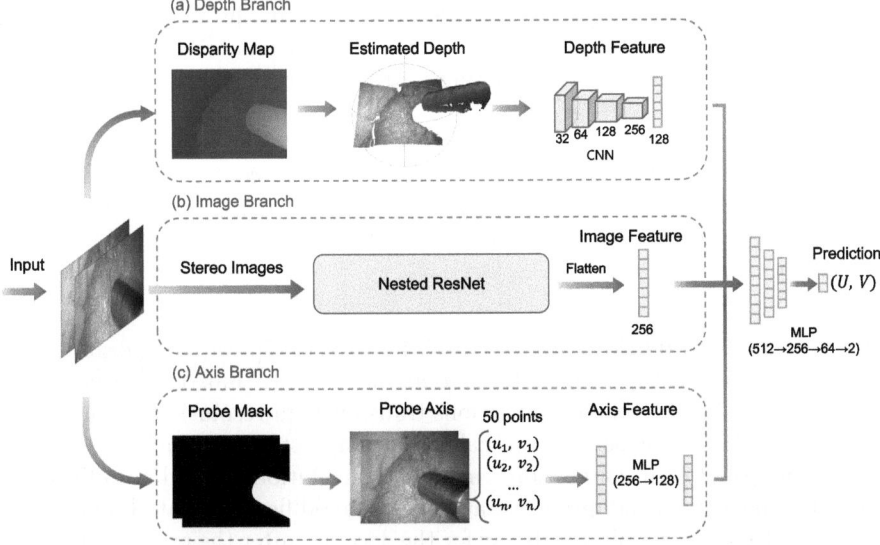

Fig. 2. The architecture for sensing area detection: stereo laparoscopic images are input on the left; branch (a) shows the depth feature extraction from depth maps using CNN layers; the main branch (b) uses a Nested ResNet for extracting features from the RGB images; branch (c) shows the feature extraction from axis points using MLP. Features from these three branches are concatenated for predicting the sensing location.

relationship of the probe and the background. To address this, we developed a Nested ResNet architecture for the image branch to extract features from the stereo laparoscopic images.

Residual Encoding Modules. The left and right images are concatenated into a 6-channel ($C = 6$) image of size $H \times W$. A 2D convolution with max pooling halves the spatial dimensions and increases the channels to $C = 64$. The resulting feature map is then fed into four residual encoding modules (Fig. 3a), each comprising one expanded bottleneck block and multiple standard bottleneck blocks.

The standard bottleneck residual block structure is widely used in classic networks such as ResNet-50, ResNet-101, and ResNet-152 [6]. In a standard bottleneck residual block (Fig. 3 (b)) we apply a 1×1 convolutional block for reducing the number of channels to $C/4$ for fast computation. After batch normalization and ELU activation [2], the middle 3×3 convolutional block is followed. Finally, a 1×1 convolutional layer recovers the channel number to the input size and an identity mapping is applied.

In addition to the standard bottleneck block, we introduce an expanded bottleneck block that increases channel depth to enhance feature representation. Unlike the standard block, which maintains channel count for identity mapping, the expanded block defers channel compression to the decoder. As shown in

Fig. 3 (c), it first reduces the channel count to $C/2$ in the middle layer, then expands it to $4C$ in the final 1×1 convolution block. To match this increase, the skip connection replaces identity mapping with a 1×1 convolution that projects the input from C to $4C$.

Using the two types of residual blocks, we construct four downsampling modules in the Nested ResNet (Fig. 3 (a)). Each module contains one expanded bottleneck block for downsampling and channel expansion. In addition, they include 3, 4, 6, and 3 standard bottleneck blocks, respectively, to increase depth and enhance feature extraction.

Global Skip Connections. Inspired by the U-Net [17], we incorporate global skip connections in the Nested ResNet architecture to link the encoder and decoder paths (Fig. 3). These connections transmit high-resolution features from early to later layers, preserving critical textural and structural information.

Within global skip connections, a 1×1 convolution is used to adjust the channel number, while adaptive average pooling modifies the spatial dimensions to suit the decoder's requirements. As for the decoder, after the residual encoding modules, partial upsampling is applied. Each upsampling block only doubles the spatial size to save computing expenses. This configuration allows high-resolution details to bypass multiple intermediate downsampling stages, using only two convolutional layers in the global skip connections applied to preserve more comprehensive information.

2.4 Loss Function

We use the square of the Euclidean distance between the ground truth and prediction as the loss function:

$$\mathcal{L} = (p_u - g_u)^2 + (p_v - g_v)^2 \tag{1}$$

where (p_u, p_v) is the predicted location and (g_u, g_v) is the ground truth value.

3 Experiments and Results

3.1 Dataset

The Coffbea dataset [8], as shown in Fig. 4, was used in this research. A silicone phantom of size $30 \times 21 \times 8 cm$ that resembled abdominal organs was used (Fig. 4 (a)). This was placed on a rotation platform alongside a fabricated SENSEI® gamma probe, with the proportion of the probe in the field of view varied between 50% and 100%. A stereo laparoscopic camera was used to capture the images. During each sampling cycle, images were captured at 10 positions, with the platform rotating by 36°C between each capture. At the end of each cycle, the relative positions of the probe and phantom to the camera were adjusted to ensure distinct views. 1200 samples were captured in total. The fabricated probe had an identical shell to a standard SENSEI® probe but was modified to include

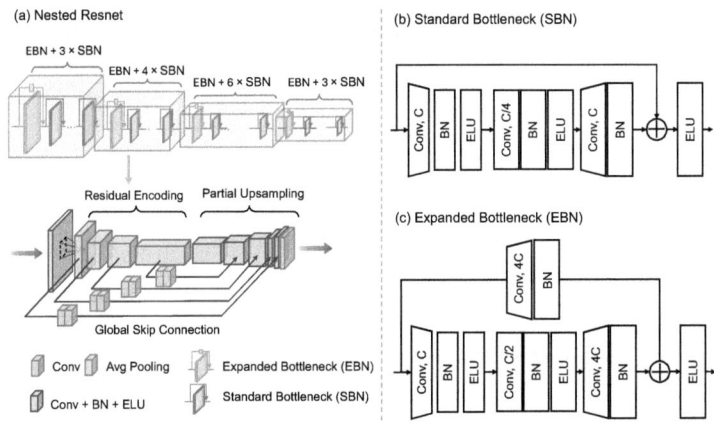

Fig. 3. The Nested ResNet design: (a) The overall architecture. (b-c) The standard bottleneck block (SBN) and the expanded bottleneck block (EBN) that form the residual encoding part in the Nested ResNet. The SBN keeps the same channel number C while the EBN expands the channel number from C to $4C$.

a laser module embedded in its casing, which was used to mark the center of the ground truth sensing area (Fig. 4 (c)). 3D maps were also included in the dataset. The 3D ground truth was generated by projecting binary-encoded structured light patterns onto the phantom, providing reliable features for stereo matching. Depth information for each point was then calculated based on the disparity and camera parameters. Some areas lack depth information due to strong reflections on the probe's metallic surface and shadowing; however, these regions lie outside the sensing area and do not affect its evaluation (Fig. 4 (d)).

3.2 Experimental Settings

We divided the 1200 images in the Coffbea dataset [8] into 800 training images, 200 validation images, and 200 test images, while ensuring images for each set were derived from different sampling cycles to avoid overlapping views. Each image was 920×1224 in size and was padded to 1224×1224. Our solution was implemented with PyTorch on an Ubuntu operating system. Experiments were conducted on an NVIDIA A40 GPU with 48 GB memory, a batch size of 8, and an Adam optimizer [10] with an initial learning rate of 1×10^{-4}, decaying to 8×10^{-5}. We trained the network for 300 epochs.

3.3 Evaluation Metrics

We calculated the Euclidean distances in the 2D space between the predicted sensing area locations and the ground truth values.

$$L_{2d} = \sqrt{(u_{\text{pred}} - u_{\text{gt}})^2 + (v_{\text{pred}} - v_{\text{gt}})^2} \qquad (2)$$

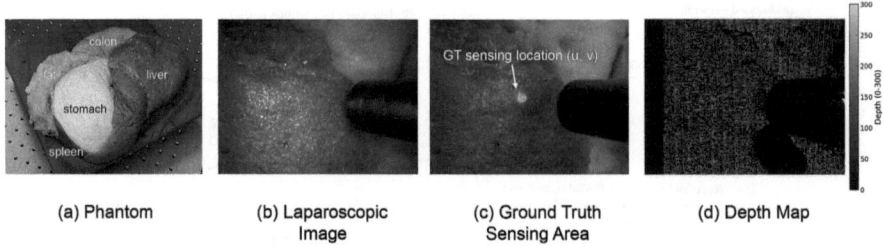

Fig. 4. The Coffbea [8] dataset: (a) the phantom of abdominal organs used for dataset collection, (b) RGB laparoscopic image, (c) ground truth sensing area provided in 2D coordinates, and (d) pre-generated depth maps.

where (u, v) is the location in 2D plane and the distance is measured in pixels. The mean error, standard deviation, and median error were reported.

Moreover, the corresponding depth information of any 2D point can be extracted from the depth map. Subsequently, both predicted and ground-truth 3D values can be obtained by back-projecting their 2D coordinates to 3D using the depth value and the camera's intrinsic parameters:

$$X = \frac{(u - o_x)Z}{\alpha}, \quad Y = \frac{(v - o_y)Z}{\beta}, \quad Z = Z \tag{3}$$

where (u, v) is the location in the 2D coordinate system, and Z is the depth value acquired from the depth map; α and β are the scaling factors in the X and Y directions; (o_x, o_y) are the coordinates of the camera's optical center. Hence, we can calculate the 3D distance:

$$L_{3d} = \sqrt{(X_{\text{pred}} - X_{\text{gt}})^2 + (Y_{\text{pred}} - Y_{\text{gt}})^2 + (Z_{\text{pred}} - Z_{\text{gt}})^2} \tag{4}$$

3.4 Quantitative Experiments

Table 1 gives the quantitative evaluation of our methods with previous solution SL Regress from Huang et al. [8]. In our solution, we utilise laparoscopic images as the primary input, from which we derive two key features: points along the probe axis and an estimated depth map. Furthermore, our solution is implemented using the Nested ResNet (NResNet) architecture proposed in this work, whereas the baseline method SL Regress [8] only combines traditional networks such as ResNet-50 [6], Vision Transformer (ViT) [4], multi-layer perceptron (MLP) and long short-term memory (LSTM) [7]. Our method outperforms all others, achieving the lowest 2D mean error with a reduction of 22.10% compared to the best results from previous studies, and reaching 43.0 pixels. The corresponding 3D mean error is also the lowest, achieving a reduction of 41.67% and reaching 3.5 mm. Given the probe's lateral resolution of 29 mm FWHM [12], this prediction error is acceptable. To build on this progress, future work will include experiments with *ex vivo* tissue samples and phantoms representing diverse surgical scenarios. This will help further evaluate and enhance the method's applicability in real-world contexts.

Table 1. Quantitative comparisons of our solution with baseline methods, SL Regress [8]. The best results achieved by our methods are shown in **bold**, while the best results from the previous method SL Regress are underlined. 2D and 3D errors are measured in pixels and millimetres respectively.

	Method			2D Metrics			3D Metrics		
	Image	Axis	Depth	Mean E.	STD	Median	Mean E.	STD	Median
SL Regress [8]	ResNet-50	-	-	85.0	65.0	66.9	10.7	15.9	6.1
	ResNet-50	LSTM	-	81.0	63.4	67.8	10.6	15.6	7.2
	ResNet-50	MLP	-	<u>55.2</u>	<u>44.7</u>	<u>40.9</u>	12.7	21.7	6.6
	ViT	-	-	66.3	65.4	49.1	8.9	15.6	4.7
	ViT	LSTM	-	73.8	68.3	57.0	<u>6.0</u>	<u>6.1</u>	<u>3.7</u>
	ViT	MLP	-	92.2	73.1	72.1	11.3	21.8	5.6
Ours	NResNet	-	-	47.8	32.5	40.4	7.6	14.0	4.1
	NResNet	MLP	-	51.5	34.4	40.0	8.7	16.6	3.2
	NResNet	-	CNN	44.4	30.6	39.6	9.2	17.0	3.7
	NResNet	MLP	CNN	**43.0**	**30.5**	**36.9**	**3.5**	**2.6**	**2.7**

3.5 Qualitative Evaluation

Visualization examples of the sensing point prediction can be seen in Fig. 5. Samples are randomly selected from the top 10%, median 10%, and worst 10% of performance based on error. The top row gives the best results from previous work [8] and the bottom row shows results using our method. It can be seen from the qualitative comparisons that our predictions achieve higher accuracy. Due to current limitations of the available dataset, we only use the dataset collected from a phantom of abdominal organs [8]. Future work will prioritise more extensive and diverse *in vitro* validation.

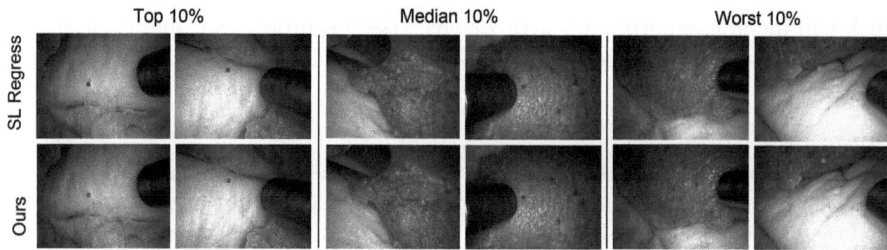

Fig. 5. Visualisation. Red dots show the location of the ground truth and green dots indicate the location of the prediction. The top row shows the results using the previous method SL Regress [8], and the bottom row shows the results of our method. (Color figure online)

4 Conclusion

We proposed a three-branch deep learning framework to improve sensing area prediction for a non imaging drop-in gamma probe. One branch uses a Nested ResNet to extract features from stereo RGB laparoscopic images, while the other two incorporate depth maps and probe axes to provide complementary geometric information. The predicted sensing location offers intuitive visual guidance to complement the audio signals of the probe. Our method significantly outperformed prior benchmarks, moving toward clinical application and reducing the difficulty of tumor localization. Future work includes validating the model with *ex vivo* tissues and animal studies within surgical workflows.

Acknowledgments. This work was supported in part by the U.K. National Institute for Health Research (NIHR) Invention for Innovation under Award NIHR200035. It is independent research funded by the National Institute for Health Research (NIHR) Imperial Biomedical Research Centre (BRC), and the Cancer Research UK (CRUK) Imperial Center.

Disclosure of Interests. The authors have no competing interests to declare that are relevant to the content of this article.

References

1. Chen, L.C., Papandreou, G., Schroff, F., Adam, H.: Rethinking atrous convolution for semantic image segmentation. arXiv preprint arXiv:1706.05587 (2017)
2. Clevert, D.A.: Fast and accurate deep network learning by exponential linear units (elus). arXiv preprint arXiv:1511.07289 (2015)
3. Dell'Oglio, P., et al.: A drop-in gamma probe for robot-assisted radioguided surgery of lymph nodes during radical prostatectomy. Eur. Urol. **79**(1), 124–132 (2021)
4. Dosovitskiy, A., et al.: An image is worth 16x16 words: transformers for image recognition at scale. In: The International Conference on Learning Representations (2021)

5. Everaerts, W., et al.: A multicentre clinical trial evaluating a drop-in gamma probe for minimally invasive sentinel lymph node dissection in prostate cancer. Eur. Urol. Focus **10**, 32–40 (2024)
6. He, K., Zhang, X., Ren, S., Sun, J.: Deep residual learning for image recognition. In: Proceedings of the IEEE Conference on Computer Vision and Pattern Recognition, pp. 770–778 (2016)
7. Hochreiter, S.: Long short-term memory. In: Neural Computation. MIT-Press (1997)
8. Huang, B., Hu, Y., Nguyen, A., Giannarou, S., Elson, D.S.: Detecting the sensing area of a laparoscopic probe in minimally invasive cancer surgery. In: Greenspan, H., et al. (eds.) MICCAI 2023. LNCS, vol. 14228, pp. 260–270. Springer, Cham (2023). https://doi.org/10.1007/978-3-031-43996-4_25
9. Junquera, J.M.A., et al.: A drop-in gamma probe for minimally invasive sentinel lymph node dissection in prostate cancer: preclinical evaluation and interim results from a multicenter clinical trial. Clin. Nuclear Med. 10–1097 (2022)
10. Kingma, D.P.: Adam: a method for stochastic optimization. arXiv preprint arXiv:1412.6980 (2014)
11. Kirillov, A., et al.: Segment anything. In: Proceedings of the IEEE/CVF International Conference on Computer Vision, pp. 4015–4026 (2023)
12. Lightpoint Medical: Miniature surgical gamma probe (2021). https://senseisurgical.com/. Accessed 19 Feb 2024
13. Maćkiewicz, A., Ratajczak, W.: Principal components analysis (PCA). Comput. Geosci. **19**, 303–342 (1993)
14. Menze, M., Geiger, A.: Object scene flow for autonomous vehicles. In: Proceedings of the IEEE Conference on Computer Vision and Pattern Recognition, pp. 3061–3070 (2015)
15. Nair, V., Hinton, G.E.: Rectified linear units improve restricted Boltzmann machines. In: Proceedings of the 27th International Conference on Machine Learning (ICML-10), pp. 807–814 (2010)
16. Povoski, S.P., et al.: A comprehensive overview of radioguided surgery using gamma detection probe technology. World J. Surg. Oncol. **7**, 1–63 (2009)
17. Ronneberger, O., Fischer, P., Brox, T.: U-net: convolutional networks for biomedical image segmentation. In: Navab, N., Hornegger, J., Wells, W.M., Frangi, A.F. (eds.) MICCAI 2015, Part III. LNCS, vol. 9351, pp. 234–241. Springer, Cham (2015). https://doi.org/10.1007/978-3-319-24574-4_28
18. Scharstein, D., et al.: High-resolution stereo datasets with subpixel-accurate ground truth. In: Jiang, X., Hornegger, J., Koch, R. (eds.) GCPR 2014. LNCS, vol. 8753, pp. 31–42. Springer, Cham (2014). https://doi.org/10.1007/978-3-319-11752-2_3
19. Schops, T., et al.: A multi-view stereo benchmark with high-resolution images and multi-camera videos. In: Proceedings of the IEEE Conference on Computer Vision and Pattern Recognition, pp. 3260–3269 (2017)
20. World Health Organization: Cancer (2022). https://www.who.int/news-room/fact-sheets/detail/cancer. Accessed 29 Dec 2024
21. Xu, H., et al.: Unifying flow, stereo and depth estimation. IEEE Trans. Pattern Anal. Mach. Intell. (2023)

Video Grounded Conversation Generation for Reference Surgical Instrument Segmentation

Yihan Wang[1], Qiao Yan[1], Lihao Liu[3], Yuchen Yuan[1], Xiaowei Hu[4], Jinpeng Li[1(✉)], and Pheng-Ann Heng[1,2]

[1] Department of Computer Science and Engineering,
The Chinese University of Hong, Hong Kong, China
jpli21@cse.cuhk.edu.hk
[2] Institute of Medical Intelligence and XR, The Chinese University of Hong Kong, Hong Kong, China
[3] Amazon, Seattle, USA
[4] School of Future Technology, South China University of Technology, Guangdong, China

Abstract. Surgical instrument segmentation in videos is essential for computer-assisted interventions, enabling accurate tool identification during surgeries. However, current Grounded Conversation Generation (GCG) methods struggle with specifying the instrument of interest and capturing complex interactions in dynamic environments due to limitations in understanding intra-frame and inter-frame information. Here, we formulate a novel Video-GCG framework for improved reference surgical instrument segmentation, which combines visual data with context-aware textual descriptions. First, we develop a Temporal Dynamic Sampling (TDDS) strategy to enhance temporal-spatial feature extraction, solving the intra-frame problem. Then, we present a mask decoding strategy to refine segmentation outputs and reduce the impact of blurred or ambiguous visual information from the surrounding environment, tackling the inter-frame problem. Experimental results show that our method outperforms the state-of-the-art VIS-Net by 18.1% and 7.5% in mAP on the EndoVis-RS17&18 datasets, showcasing superior performance and efficiency with fewer computational resources. Codes will be released.

Keywords: Reference Surgical Instrument Segmentation · Video Grounded Conversation Generation · Multimodal Large Language Models

1 Introduction

Surgical video instrument segmentation plays an important role in computer-assisted interventions, with applications in endoscopic and robotic surgeries.

Supplementary Information The online version contains supplementary material available at https://doi.org/10.1007/978-3-032-09784-2_13.

Fig. 1. Comparison of different segmentation task. Our approach enhances the segmentation by taking in comprehensive target instrument descriptions, effectively distinguishing multiple instruments within a frame.

This process is fundamental for downstream tasks such as trajectory prediction [15], tool pose estimation [7,19], and robotic navigation [23]. Deep learning approaches [3,4,9,12,14,20,22,24] enhance accuracy by using feature extraction capabilities. While standard video segmentation methods aim to segment all visible instruments, they are not designed to focus on a *specific instrument of interest* or to *incorporate contextual cues* necessary for complex surgical environments.

As a result, reference segmentation has been proposed for surgical instrument segmentation. Reference segmentation aims to segment instruments specified by a textual query, where the reference often aligns with the surgeons intended focus at a given surgical stage. However, reference segmentation lacks the capability to engage in natural, meaningful conversations, *limiting its usefulness in interactive tasks* that demand a comprehensive understanding of both visual and textual information.

Grounded Conversation Generation (GCG) [17] offers a solution to overcome the limitations of reference segmentation, which generates natural language responses interleaved with object segmentation masks. As shown in Fig. 1, this approach is particularly advantageous for surgical video analysis, as it serves as a natural intermediate for interacting with surgeons via language while enhancing instruction-aware segmentation and supporting potential downstream tasks. However, GCG faces two main challenges in surgical video scenarios: **(1) Intra-frame problem**: The constantly changing surgical environment, with instruments moving, shifting targets, and overlapping tools, leads to blurred or ambiguous visual information, complicating the segmentation and localization tasks. **(2) Inter-frame problem**: Background objects, patient tissue structures, and other environmental factors can interfere with the accurate identification and localization of surgical instruments across frames, reducing the effectiveness of GCG in dynamic video settings.

In this paper, to address the challenges and inherent drawbacks of GCG in reference surgical video segmentation, we develop a novel framework called Video-based Grounded Conversation Generation (Video-GCG) for reference surgical instrument segmentation. First, to tackle the intra-frame problem of rapid and complex localization of surgical tools, we present a Temporal Dynamic

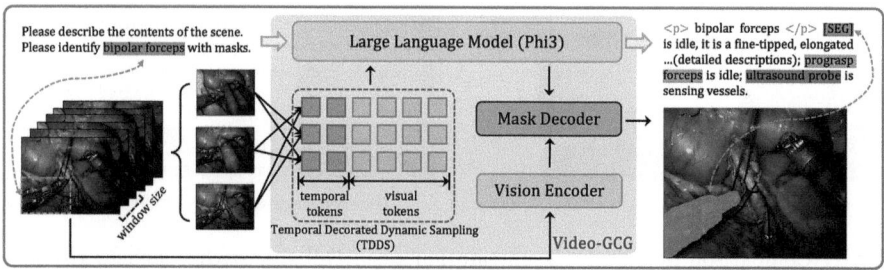

Fig. 2. Overview of our Video-GCG. The model processes video sequences and text inputs in parallel. We introduce the TDDS module to address the intra-frame problem, while the mask decoder is designed to handle the inter-frame challenges. The model outputs precise text answers and mask predictions.

Sampling (TDDS) technique. This strategy ensures that information from each frame is enriched with contextual information from its adjacent frames, thereby enhancing GCGs understanding capacity and improving video segmentation in dynamic surgical environments. Second, to address the inter-frame problem, we introduce a mask decoder from [11]. Its strong visual representation capability enables effective extraction of rich features and contextual cues, refining segmentation and supporting precise referring segmentation even in complex scenes with occlusions or close instrument interactions. Various experiments on the EndoVis-RS17&18 benchmarks demonstrate that our approach achieves comparable or superior performance on the conversational task. Furthermore, on the segmentation task, it significantly improves segmentation accuracy and robustness, outperforming existing methods in surgical video scenarios.

2 Methodology

2.1 Overview

Figure 2 shows the overall framework of our method, which processes video sequences and instruction text as parallel inputs. For the instruction understanding, the text input is first tokenized using a text tokenizer and then passed into a large language model (LLM). Simultaneously, the video sequences are processed using a sliding window with a fixed size W, where frames within each window are passed through a visual tokenizer to generate initial visual tokens. To address intra-frame challenges, we introduce the Temporal Decorated Dynamic Sampling (TDDS) module, which enhances both the temporal and spatial features of these visual tokens. For mask prediction, the current frame is processed through a mask decoder via an additional visual encoder, along with the LLM embeddings from the text branch, to address inter-frame issues. The model output consists of both text descriptions and mask predictions, which are obtained by decoding the outputs of the LLM and the mask decoder, respectively.

2.2 Reference Surgical Instrument Segmentation

GCG for Intention-Orientated Segmentation. Only scene-level descriptions are insufficient for surgical videos, as the status of different instruments within a single scene are often very similar or identical, which can lead to misinterpretations or incorrect judgments by the model.

To address this, we extend the GCG paradigm into an intention-oriented approach through two key innovations. **(1)** In [21], intention-oriented segmentation relies on a simple status description of the target instrument, such as *the bipolar forceps is idle*. This format lacks sufficient detail to distinguish the target instrument from others in the scene. To resolve this, we propose reorganizing the sequence-level target status descriptions into frame-level, comprehensive scene descriptions that include the status of all instruments appearing in the scene. **(2)** As in Fig. 1, by predicting attributes such as function, appearance, and contextual relevance of specific instruments, the model can learn more accurate grounding and segmentation of the target within the scene. To achieve this, we generate detailed descriptions for each object in the dataset, allowing the model to learn fine-grained visual-text correspondences. Specifically, we use GPT-4 [2] with a tailored prompt to produce comprehensive descriptions for each instrument, following [21]. Table 3 demonstrates the efficacy of our approach.

Once the detailed descriptions are generated, we integrate them with scene-level descriptions of the specific instrument in **(1)**. The final output consists of enriched captions that combine these detailed descriptions with interleaved segmentation masks, along with simple annotations for the other instruments in the frame. Similar to [17], we use special tokens, namely <p>, </p> and [SEG], to represent the start and end of the referred instruments name and its corresponding region mask, respectively. This format offers a comprehensive portrayal of the referred instrument, adding crucial functional context and effectively distinguishing multiple objects within the frame. By training with the fine-grained descriptions, our method not only produces segmentation masks for the targeted instrument but also provides detailed insights into its appearance and functionality, as illustrated in Fig. 1.

2.3 Video Grounded Conversation Generation (Video-GCG)

Temporal Decoration Dynamic Sampling for Intra-frame Problem. To tackle the intra-frame and inter-frame problems, existing methods often downsample frames using pooling techniques [5,10,13] to accelerate visual token generation. However, as discussed in Sect. 1, in robotic surgery, medical instruments move rapidly in surgical videos. These approaches risk losing crucial visual information, potentially resulting in a complete loss of context in certain frames.

To address this issue, we propose a new sampling method called Temporal Decoration Dynamic Sampling (TDDS) strategy for surgical video frame processing. First, the video sequence L is divided into N segments, each of window size W, such that $L = N \times W$. If the length of the video sequence cannot be evenly divided by W, the remaining frames are repeated to form a complete window. Specifically, as illustrated in Fig. 2, within a sliding window, we incorporate features from the two adjacent frames as additional temporal tokens. This not only preserves the full resolution of each frame but also enriches the feature set with temporal information derived from the neighboring frames. By concatenating two neighboring temporal tokens with the features of the current frame, we create a more comprehensive representation that captures the rapid changes occurring in the surgical environment. As shown in Table 3, in comparison to the method outlined in [5], our approach significantly enhances critical temporal and spatial information while incurring minimal computational overhead, adding only two extra tokens per frame. This efficient balance enables us to optimize the input to the LLM, allowing the model to preserve spatial detail and temporal context effectively, resulting in more accurate segmentation for complex surgical video processing.

Mask Decoder for Inter-frame Problem. We take [11] as our promptable mask decoder to address the inter-frame problem. It predicts masks and their category probabilities based on the internal output features of the vision encoder and the LLM output text embeddings corresponding to the [SEG] token. This mask decoder is guided by the [SEG] token embedding derived from the LLM, which serves as a consistent text prompt across frames. This stabilizes mask prediction by anchoring the decoders attention to the same object category (*e.g.*, grasping forceps), regardless of appearance changes or background clutter.

2.4 Training Objective

The model is trained end-to-end using the text generation cross-entropy loss $\mathcal{L}_{text}$ and segmentation loss, which consists of DICE loss $\mathcal{L}_{dice}$ and a per-pixel class-rebalanced binary cross-entropy (CRCE) loss $\mathcal{L}_{\text{CRCE}}$ as shown in Eq. 1 and Eq. 2, where w_{cb} is the class-balanced weight designed to address the class imbalance issue in the EndoVis-RS17&18 datasets. The hyperparameter $\beta \in [0, 1)$ controls how fast w_{cb} grows as the number of samples n_c increases.

$$w_{cb} = log[1 + (\frac{1-\beta}{1-\beta^{n_c}} + 1e^{-6})] \qquad (1)$$

$$\mathcal{L}_{\text{CRCE}} = F_{scale} \cdot \frac{w_{cb} + f}{\sum_{n_c}(w_{cb} + f)} \cdot log(\frac{\exp(z_c)}{\sum_{i=1}^{c} \exp(z_i)}) \qquad (2)$$

where f is a Laplace smoothing factor, z_i is a specific class from all classes z_c.
The final loss is computed as the weighted sum of the three losses:

$$\mathcal{L}_{total} = w_t \cdot \mathcal{L}_{txt} + w_m \cdot (w_d \cdot \mathcal{L}_{dice} + w_c \cdot \mathcal{L}_{\text{CRCE}}) \qquad (3)$$

3 Experiments and Discussion

3.1 Dataset, Evaluation Metrics and Implementation Details

Dataset and Evaluation Metrics. We train and validate our method on the EndoVis RS-17 and EndoVis RS-18 datasets [21]. For Video-based GCG task,

Table 1. Performance on Video-based GCG task: Metrics include METEOR (M), Rouge-L(R-L), Bleu-4 (B-4), AP50, mIoU, and Mask Recall. * denotes finetuned on the EndoVis-RS17 dataset or the EndoVis-RS18 dataset.

Method	Param.	EndoVis-RS17						EndoVis-RS18					
		AP50	mIoU	Recall	M	R-L	B-4	AP50	mIoU	Recall	M	R-L	B-4
LISA*	7B	70.3	75.8	83.7	55.7	80.7	67.1	43.7	58.9	66.1	51.2	77.8	**58.8**
GLAMM*	7B	70.7	76.0	84.1	**56.6**	**81.5**	**68.5**	46.5	61.9	68.2	52.3	77.2	58.4
Ours	3.8B	**76.8**	**79.4**	**87.5**	55.1	79.0	66.0	**64.1**	**73.0**	**80.1**	52.6	77.8	58.3

we adopt the evaluation metrics from [17], which assess four key aspects: (i) generated caption quality, (ii) mask-to-phrase alignment accuracy, (iii) generated mask quality, and (iv) region-specific grounding ability. Metrics include METEOR, Rouge-L, and Bert-4 for caption quality, class-agnostic mask Average Precision (AP50) for alignment, mean mask Intersection over Union (mIoU) for segmentation, and mask recall (Recall) for region-specific grounding. AP50 is computed for an IoU greater than or equal to 0.50, while mAP is calculated for the IoU range from 0.50 to 0.95 with an interval of 0.05. For the downstream instrument segmentation task, we use widely adopted metrics from [21], including Precision@K, Overall IoU (O. IoU), Mean IoU (M. IoU), and mAP.

Implementation Details. We implement our model using LLaVA-Phi-3-V [16], based on the 3.8B Phi-3 model [1], and adopt the vision encoder and mask decoder from SAM [11]. Training is performed on EndoVis-RS17/18 with window size $W = 3$, using random flip and crop for augmentation. The model is trained for 100 epochs on 8A100 (80GB) GPUs with DeepSpeed [18], using AdamW (lr=1e-4, no weight decay) and WarmupDecayLR (100 warmup steps). We set $w_t = w_m = 1.0$, $w_d = 0.5$, and $w_c = 2.0$, with a batch size of 2 per device. Total training time is about 2 h.

For EndoVis-RS17, we follow the six-fold cross-validation approach used by VIS-Net to replicate their baseline performance [21]. For EndoVis-RS18, we follow the train-validation split in [21]. In Table 2, other Vision Language Models (VLMs) are reproduced based on our data split, while other specialist models adhere to the performance metrics reported in [21].

3.2 Experimental Results

Comparison with State-of-the-Art (SoTA). We compare our method with current SoTA VLM-based approaches on the Video-based GCG task using the EndoVis-RS17&18 datasets. The evaluation results presented in Table 1, show that our method demonstrates competitive performance. Notably, among VLM-based models with reasoning capabilities, despite our model having only 3.8B parameters, outperforms larger models such as LISA-7B and GLAMM-7B. Furthermore, we evaluate our approach against both segmentation specialists and VLMs on the downstream segmentation task, as shown in Table 2. Our method

Table 2. Performance on downstream segmentation task. * denotes finetuned on the EndoVis-RS17 or the EndoVis-RS18 dataset. # denotes the output descriptions.

Dataset	Model Category	Method	Precision					IoU		mAP
			P@0.5	P@0.6	P@0.7	P@0.8	P@0.9	O.IoU	M.IoU	
EndoVis -RS17	Specialist	LBDT [8]	71.0	64.5	60.1	52.5	29.6	61.0	65.6	51.9
		MTTR [6]	71.7	64.8	60.9	52.9	30.2	61.3	66.3	52.2
		VIS-Net [21]	76.1	66.4	62.3	54.6	32.2	65.5	70.1	53.8
	VLM	LISA*	83.7	81.0	77.1	70.4	48.7	72.9	75.8	67.6
		GLAMM*	84.1	81.7	78.0	72.3	46.8	72.9	76.0	68.0
		VideoLISA*	85.4	83.4	79.7	72.9	49.9	76.0	77.4	69.8
		Ours(w/o #)	87.1	85.1	81.6	73.9	50.2	78.0	78.7	71.0
		Ours	**87.5**	**85.5**	**82.0**	**75.0**	**52.7**	**78.8**	**79.4**	**71.9**
EndoVis -RS18	Specialist	LBDT [8]	76.5	72.0	65.7	56.7	34.0	71.9	70.2	57.5
		MTTR [6]	77.1	72.0	66.0	57.6	34.8	72.2	70.8	58.0
		VIS-Net [21]	**80.2**	75.2	68.5	60.2	36.1	**74.2**	72.3	60.1
	VLM	LISA*	66.1	63.9	58.2	49.6	31.0	47.9	58.9	50.1
		GLAMM*	68.2	66.3	62.5	57.2	41.2	53.1	61.9	55.6
		VideoLISA*	73.7	72.0	68.2	61.4	43.5	65.9	67.0	60.2
		Ours(w/o #)	73.6	72.1	70.1	66.0	50.8	65.7	68.0	62.8
		Ours	80.1	**78.4**	**76.0**	**70.2**	**52.2**	74.1	**73.0**	**67.6**

Table 3. Ablation study of the influence of text quality and sampling strategy. ‡[SEG]: with interleaved [SEG] token.

Influence of Text Quality				Influence of Sampling Strategy			
Method	O. IoU	M. IoU	mAP	Method	O. IoU	M. IoU	mAP
only [SEG]	65.7	68.0	62.8	nFrame	67.4	68.3	60.9
[SEG]+ gen-desc	68.8	68.6	62.7	Slow-Fast Pooling [10]	68.9	65.6	59.1
gen-desc + ‡[SEG]	66.9	67.0	61.4	Sparse-Dense Sampling [5]	70.1	69.8	61.6
Ours Video-GCG	**74.1**	**73.0**	**67.6**	Ours TDDS	**74.1**	**73.0**	**67.6**

consistently outperforms both specialist and VLM-based models across both datasets, highlighting its effectiveness in intention-oriented segmentation.

Ablation Studies. We conduct extensive ablation studies on the EndoVis-RS18 dataset to evaluate various design choices of our model. **(a) Analysis of the Influence of Text Quality.** We conduct an ablation study to assess the impact of different text qualities on segmentation performance. Four experiments are performed: (1) using only the [SEG] special token, (2) using a separate [SEG] token with a general description, (3) combining the general description with an interleaved [SEG] token, and (4) combining the general description with intention-oriented fine-grained descriptions and an interleaved [SEG] token

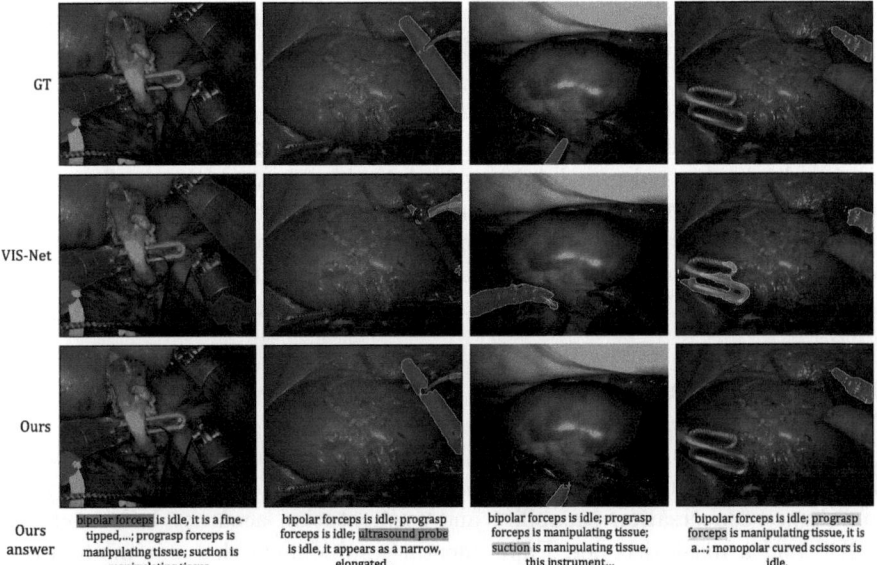

Fig. 3. Qualitative Analysis on EndoVis-RS18 dataset. Our model performs better by providing more accurate locations of referred objects, with grounded context that offers additional clues for differentiating between various surgical instruments. More visualization results can be found in the supplementary material.

Table 4. Performance under different Window Sizes

W	P@0.5	P@0.6	P@0.7	P@0.8	P@0.9	O. IOU	M. IOU	mAP
1	77.9	76.2	72.5	65.9	48.6	73.0	70.3	64.4
3	80.3	78.1	75.2	68.8	50.5	74.9	72.9	66.6
5	77.3	75.7	73.4	67.1	48.9	73.2	70.3	64.9

(ours). The left side of Table 3 shows that additional text descriptions improve segmentation performance (comparison between (1) and (2)). However, simple descriptions of intention-oriented target objects can negatively affect segmentation accuracy (performance drop from (2) to (3)), as similar state descriptions are applied to different objects, causing confusion and reducing precision. In contrast, the results of (4) demonstrate that high-quality fine-grained descriptions significantly enhance segmentation performance. **(b) Analysis of the Influence of Sampling Strategy.** On the right side of Table 3, we compare our proposed TDDS module with several alternative design choices. These approaches include: (1) nFrame, which directly concatenates visual tokens from multiple frames; (2) Slow-Fast Pooling, which pools each frame at varying strengths in a slow-fast manner [10]; and (3) Sparse-Dense Sampling, which randomly selects sparse frames and reduces them to a single visual token [5]. The experimental results show that our TDDS consistently outperforms other VLMs' sampling

methods. Furthermore, to assess the impact of different window sizes, we conduct an ablation study as shown in Table 4. The results show that using a small window size ($W = 1$) restricts the model to the current frame, making it prone to errors under occlusion, blur, or fast motion due to missing temporal context. In contrast, a large window size ($W = 5$) introduces excessive temporal information, leading to noise, redundancy, and motion blur, which can distract the model from focusing on relevant regions in the current frame. Therefore, both too small and too large window size make the intra-frame problem more serious.
Visual Comparison. The comparison of the segmentation results between our model and one SoTA segmentation specialist VIS-Net [21] is shown in Fig. 3.

4 Conclusion

In this paper, we propose Video-GCG, a novel framework for reference surgical instrument segmentation. We solve both the intra-frame and inter-frame problem to enhance conversation capability and segmentation accuracy by introducing TDDS and a mask decoder. Experimental results show that our method outperforms existing approaches, achieving the state-of-the-art performance. This work advances efficient, context-aware segmentation for surgical video analysis.

Acknowledgment. The work described in this paper was supported by the Research Grants Council of the Hong Kong Special Administrative Region, China, in part under Project T45-401/22-N and in part under Project AoE/E-407/24-N.

References

1. Abdin, M., et al.: Phi-3 technical report: a highly capable language model locally on your phone. arXiv preprint arXiv:2404.14219 (2024)
2. Achiam, J., et al.: GPT-4 technical report. arXiv preprint arXiv:2303.08774 (2023)
3. Ayobi, N., Pérez-Rondón, A., Rodríguez, S., Arbeláez, P.: MATIS: masked-attention transformers for surgical instrument segmentation. In: 2023 IEEE 20th International Symposium on Biomedical Imaging (ISBI), pp. 1–5. IEEE (2023)
4. Baby, B., et al.: From forks to forceps: a new framework for instance segmentation of surgical instruments. In: Proceedings of the IEEE/CVF Winter Conference on Applications of Computer Vision, pp. 6191–6201 (2023)
5. Bai, Z., et al.: One token to seg them all: language instructed reasoning segmentation in videos. arXiv preprint arXiv:2409.19603 (2024)
6. Botach, A., Zheltonozhskii, E., Baskin, C.: End-to-end referring video object segmentation with multimodal transformers. In: Proceedings of the IEEE Conference on Computer Vision and Pattern Recognition (CVPR) (2022)

7. Cheng, Y., Liu, L., Wang, S., Jin, Y., Schönlieb, C.B., Aviles-Rivero, A.I.: Why deep surgical models fail?: Revisiting surgical action triplet recognition through the lens of robustness. In: International Workshop on Trustworthy Machine Learning for Healthcare, pp. 177–189. Springer (2023)
8. Ding, Z., Hui, T., Huang, J., Wei, X., Han, J., Liu, S.: Language-bridged spatial-temporal interaction for referring video object segmentation. In: Proceedings of the IEEE/CVF Conference on Computer Vision and Pattern Recognition, pp. 4964–4973 (2022)
9. González, C., Bravo-Sánchez, L., Arbelaez, P.: ISINet: an instance-based approach for surgical instrument segmentation. In: International Conference on Medical Image Computing and Computer-Assisted Intervention, pp. 595–605. Springer (2020)
10. Huang, D.A., et al.: Lita: Language instructed temporal-localization assistant. In: European Conference on Computer Vision, pp. 202–218. Springer (2024)
11. Kirillov, A., et al.: Segment anything. In: Proceedings of the IEEE/CVF International Conference on Computer Vision, pp. 4015–4026 (2023)
12. Li, P., Liu, L., Schönlieb, C.B., Aviles-Rivero, A.I.: Optimised propainter for video diminished reality inpainting. arXiv preprint arXiv:2406.02287 (2024)
13. Maaz, M., Rasheed, H., Khan, S., Khan, F.S.: Video-ChatGPT: towards detailed video understanding via large vision and language models. arXiv preprint arXiv:2306.05424 (2023)
14. Ni, Z.L., Bian, G.B., Hou, Z.G., Zhou, X.H., Xie, X.L., Li, Z.: Attention-guided lightweight network for real-time segmentation of robotic surgical instruments. In: 2020 IEEE International Conference on Robotics and Automation (ICRA), pp. 9939–9945 (2020). https://doi.org/10.1109/ICRA40945.2020.9197425
15. Osa, T., Sugita, N., Mitsuishi, M.: Online trajectory planning and force control for automation of surgical tasks. IEEE Trans. Autom. Sci. Eng. **15**(2), 675–691 (2018). https://doi.org/10.1109/TASE.2017.2676018
16. Rasheed, H., Maaz, M., Khan, S., Khan, F.S.: LLaVA++: extending visual capabilities with LLaMA-3 and Phi-3 (2024). https://github.com/mbzuai-oryx/LLaVA-pp
17. Rasheed, H., et al.: GLaMM: pixel grounding large multimodal model. In: Proceedings of the IEEE/CVF Conference on Computer Vision and Pattern Recognition, pp. 13009–13018 (2024)
18. Rasley, J., Rajbhandari, S., Ruwase, O., He, Y.: DeepSpeed: system optimizations enable training deep learning models with over 100 billion parameters. In: Proceedings of the 26th ACM SIGKDD International Conference on Knowledge Discovery & Data Mining, pp. 3505–3506 (2020)
19. Sarikaya, D., Corso, J.J., Guru, K.A.: Detection and localization of robotic tools in robot-assisted surgery videos using deep neural networks for region proposal and detection. IEEE Trans. Med. Imaging **36**(7), 1542–1549 (2017). https://doi.org/10.1109/TMI.2017.2665671
20. Shvets, A.A., Rakhlin, A., Kalinin, A.A., Iglovikov, V.I.: Automatic instrument segmentation in robot-assisted surgery using deep learning. In: 2018 17th IEEE International Conference on Machine Learning and Applications (ICMLA), pp. 624–628 (2018). https://doi.org/10.1109/ICMLA.2018.00100
21. Wang, H., et al.: Video-instrument synergistic network for referring video instrument segmentation in robotic surgery. IEEE Trans. Med. Imaging (2024)
22. Yue, W., Zhang, J., Hu, K., Xia, Y., Luo, J., Wang, Z.: SurgicalSAM: efficient class promptable surgical instrument segmentation (2024)

23. Zang, D., Bian, G.-B., Wang, Y., Li, Z.: An extremely fast and precise convolutional neural network for recognition and localization of cataract surgical tools. In: Shen, D., et al. (eds.) MICCAI 2019. LNCS, vol. 11768, pp. 56–64. Springer, Cham (2019). https://doi.org/10.1007/978-3-030-32254-0_7
24. Zhao, Z., Jin, Y., Gao, X., Dou, Q., Heng, P.A.: Learning motion flows for semi-supervised instrument segmentation from robotic surgical video. In: Medical Image Computing and Computer Assisted Intervention–MICCAI 2020: 23rd International Conference, Lima, Peru, October 4–8, 2020, Proceedings, Part III 23, pp. 679–689. Springer (2020)

Multi-stage CNN for Fast Registration of 3D Preoperative CTs to 2D Intraoperative X-Rays

Federica Facente[1](✉), Benjamin Billot[1], Vivek Gopalakrishnan[3], Manasi Kattel[1], Wen Wei[2], Polina Golland[3], Hervé Delingette[1], Nicholas Ayache[1], and Pierre Berthet-Rayne[2]

[1] Université Côte d'Azur, Inria, Epione Team, Sophia Antipolis, France
federica.facente@inria.fr
[2] Caranx Medical, Nice, France
[3] CSAIL, MIT, Cambridge, USA

Abstract. Minimally invasive interventions often rely on live 2D X-rays for image guidance. Yet, anatomical localization and procedural accuracy can be enhanced by spatial alignment of these intraoperative X-rays with 3D preoperative computed tomographies (CTs). This 3D/2D registration problem is typically formulated as pose estimation of the X-ray source relatively to the CT, which is done by simulating synthetic X-rays from 2D projections of CT volumes. However, the optimization-based refinement used by the state-of-the-art deep learning approach takes several seconds, thus exceeding the allowed time budget in live image guidance. In this paper, we propose LXPose (Live X-ray Pose estimation), a self-supervised multi-stage 3D/2D registration framework for real-time image guidance. LXpose removes the dependency on optimization and leverages a two-stage CNN trained with a projection loss to ensure high accuracy and computational efficiency. Moreover, we apply extensive data augmentation to mitigate the domain gap between simulated and real X-rays. Overall, LXPose yields comparable 2D registration error to the state-of-the-art method, while reducing inference time to 20 ms, which demonstrates the potential of LXPose for real-time clinical deployment. Our code is available at https://github.com/fedefacente/LXPose.git.

Keywords: Pose estimation · Multi-stage CNN · Image guidance

1 Introduction

Minimally invasive interventions are often guided using real-time X-rays, which are acquired with a mobile C-arm device. However, X-ray imaging inherently lacks depth information since it is a 2D projection of 3D anatomy. For this reason, intervention planning is often performed on preoperative 3D computed tomographies (CTs) of high resolution and superior soft tissue contrast. CT-derived preoperative information is then mentally projected by the clinician onto the intraoperative X-rays, which increases cognitive overload and risk of errors [30,36]. A solution to this problem is to automatically register the CT volumes onto the live X-ray images, thus effortlessly providing the clinician with preoperative information. Nonetheless, CT to X-ray registration remains a major challenge due to the complexity of the 3D/2D mapping.

Early methods are based on optimization, and use the fact that X-rays and CTs share the same imaging physics. Indeed, X-rays can be realistically approximated as 2D projections of 3D CTs. These projections, known as digitally reconstructed radiographs (DRR), are simulated by virtually casting rays from the X-ray source (i.e., the C-arm) and computing the attenuation along ray propagation through the CT [21,24]. As a result, optimization strategies formulate the CT/X-ray registration task as pose estimation of the C-arm relatively to the CT volume, so that preoperative information can easily be projected onto intraoperative X-rays. Specifically, pose prediction is performed by optimizing a similarity metric between pose-dependent DRRs and the real X-rays [1,13,25]. While a more recent simulation framework (DeepDRR [33]) improves the accuracy of optimization methods [15], these remain sensitive to local minima.

Modern CT/X-ray registration methods mostly rely on deep learning. A first class of methods performs C-arm pose estimation by training supervised X-ray landmark detection networks in either cross-subject settings (for generalizability) [2,27,34] or subject-specific scenarios (for enhanced accuracy) [6,12,14]. The detected landmarks are then matched to 3D CT keypoints via Perspective-n-Point (PnP) [16] for final pose estimation. However, these methods require labor-intensive manual annotations of matched 3D/2D landmarks. Recent approaches propose to reduce this dependency by regressing 2D projections of dense surface coordinates obtained from CT segmentations [18,31], but such methods remain sensitive to occlusions by surgical tools. Another class of methods involves directly regressing the C-arm pose from input X-rays using patient-specific models trained on simulated DRRs [3,23,26] possibly along with real X-rays paired with cumbersome ground truth poses [8,38]. More generally, predicting accurate poses in only one forward pass remains challenging. That is why, state-of-the-art methods propose to treat the poses regressed by networks as initialization for optimization-based refinement [9,10,37]. Importantly, these methods introduce differentiable variants of DeepDRR, namely DiffDRR [11] and diffDeepDRR [37], to enable gradient-based optimization. However, this optimization step takes several seconds, which far exceeds the 40 ms constraint imposed by the maximum acquisition rate of X-ray systems used for real-time image guidance. Finally, all previous methods suffer from the substantial domain gap between synthetic DRRs and real X-rays [4,32].

Here, we propose LXPose (Live X-ray Pose estimation), a novel real-time multi-stage patient-specific convolutional neural network (CNN) for fast and accurate pose prediction. Specifically, we (i) employ a two-stage CNN that progressively refines pose estimates while harnessing the speed of a single forward

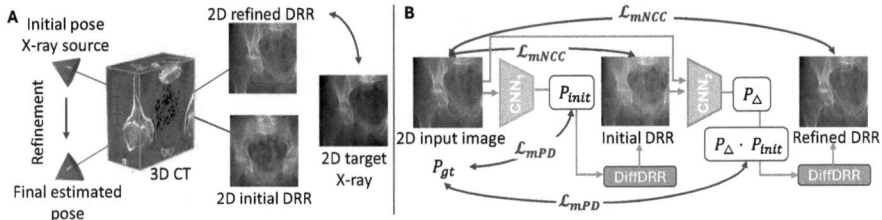

Fig. 1. (A) Method overview. LXPose performs CT/X-ray registration by progressively estimating the pose of the X-ray source from which 2D projections of the CT (i.e., DRRs) best align with the input X-ray. (B) LXPose architecture. A first CNN regresses the pose parameters from a 2D input image. This initial pose is then used to simulate a corresponding DRR. Finally, the latter is used by a second CNN along with the input image to predict a corrective term for final pose estimation.

pass for real-time deployment; (*ii*) minimize a mean Projection Distance (mPD) training loss to directly optimize alignment in the space of 2D X-rays used during intervention; (*iii*) precisely quantify the domain gap between real X-rays and DRRs, and partly address it with specific data augmentations. Overall, our method achieves comparable 2D projection error as the state-of-the-art methods, while enabling real-time integration of CT-based information during intervention by considerably reducing inference time to 20 ms.

2 Methods

Our goal is to register 3D preoperative CT volumes to 2D intraoperative X-rays. This is achieved by training a multi-stage CNN for progressive pose estimation of the C-arm in the coordinate system of the CT scanner. In line with the literature, LXPose is trained using synthetic DRRs simulated by casting virtual rays through the preoperative CT [9,37]. Our method is illustrated in Fig. 1.

2.1 X-Ray Projective Geometry

The C-arm is a mobile system with six degrees of freedom (6DoF). Its pose is defined by a translation $t = (t_x, t_y, t_z) \in \mathbb{R}^3$ controlling the in-plane position (t_x, t_y) and magnification t_z, as well as by a rotation $r = (\alpha, \beta, \gamma) \in \mathbb{R}^3$, where α and β are the orbital and secondary angles in the plane of the C-arm gantry, and γ is the angle around the normal of the detector.

Building on previous works, we model the geometric projection of CT volumes to 2D DRRs using the pinhole camera model [16], which describes how 3D points are projected onto a 2D image. This model relies on a $\mathbb{R}^{3\times 4}$ projection matrix Π that maps 3D world points in homogeneous coordinates $(x, y, z, 1)$ onto 2D pixels homogeneous coordinates $(u, v, 1)$. This projection matrix Π can be decomposed into extrinsic and intrinsic matrices. The former, which is at the core of this paper, is the $\mathbb{R}^{4\times 4}$ matrix corresponding to the rigid transform that converts world coordinates (here the CT scanner coordinates) to the camera coordinate system. Thus, the extrinsic matrix is equivalent to the pose P of the C-arm, and is fully described by $[r,t]$. Then, the intrinsic component is a $\mathbb{R}^{3\times 4}$ matrix that transforms points from the camera coordinate system to pixel locations in image space (i.e., the 2D X-ray), and depends on the camera properties: focal length f, pixel size, and image center.

Overall, given a calibrated C-arm system with known intrinsic properties, the 2D projection X^{2D} of a 3D point X^{3D} in homogeneous coordinates is fully determined by the pose P of the C-arm: $X^{2D} = \Pi_P(X^{3D})$.

2.2 Synthetic X-Ray Simulation

While not being the focus of this paper, we summarize here the principles of DRR simulation, and refer the reader to [11,37] for more detailed descriptions. Briefly, DRR simulation frameworks model the attenuation undergone by X-rays when travelling through tissues [24]. This model builds on the fact that CTs are acquired using the same imaging physics as X-rays [21]. Specifically, DRRs are simulated by casting rays from a given position of the C-arm through 3D CTs. Pixel values of the DRR are then obtained by simulating the intensity of the outgoing rays, which linearly depends on attenuation coefficients defined by the density of the traversed tissues [33].

2.3 Multi-stage CNN

LXPose is a two-stage CNN framework that takes a 2D X-ray as input and predicts the corresponding 6DoF C-arm pose in the space of the 3D CT. This is achieved by using a first network CNN1 to regress an initial estimate of $[r,t] \in \mathbb{R}^6$ (Fig. 1.B). Here, r is modelled with an axis-angle parametrization, which slightly improved results in preliminary experiments compared to Euler angles used in previous works [9]. A corresponding $\mathbb{R}^{4\times 4}$ representation of this initial predicted pose P_{init} is then constructed using the Rodrigues Formula [16].

To enhance robustness and accuracy, a second network CNN2 refines the initial prediction. Specifically, CNN2 receives both the input X-ray and a DRR simulated from the first pose estimate (initial DRR in Fig. 1.B), and predicts a corrective transform P_Δ, which is composed to P_{init} to obtain a final pose $P_{\text{final}} = P_\Delta \cdot P_{\text{initial}}$. Overall, CNN2 replaces the time-consuming optimization used in state-of-the-art methods [9,10,37], thus providing a computationally efficient alternative suitable for real-time clinical use.

2.4 Learning

Building on the CT/X-ray deep learning registration literature, we train LXPose using simulated DRRs, which suppresses the need for ground truth poses or matched 3D/2D landmarks annotated on real data. Here, we propose to use an anatomical supervision that relies only on a set of N CT-derived 3D anatomical landmarks $\{X_n^{3D}\}_{1 \leq n \leq N}$. Specifically, we optimize the mean Projection Distance (mPD) loss, defined as the average Euclidean distance between the 2D projections of the 3D landmarks using the ground truth and predicted poses:

$$\mathcal{L}_{\mathrm{mPD}} = \frac{1}{N} \sum_{n=1}^{N} \left\| \Pi_{P_{\mathrm{GT}}}(X_n^{3D}) - \Pi_{P_{\mathrm{final}}}(X_n^{3D}) \right\|_2, \qquad (1)$$

where P_{GT} is the ground truth C-arm pose. For clinical interpretability, the mPD is reported in millimeters using the known pixel size of the X-ray system. Moreover, during the loss computation, we discard landmarks whose 2D ground truth falls outside the field of view of the reference DRR, which slightly improved results in preliminary experiments by allowing the network to focus only on visible projected points. More generally, we emphasize that, as opposed to state-of-the-art methods that employ geometric losses directly defined on the regressed poses [9,37], our network is trained by optimization in the 2D projection space, which better reflects the clinical scenario of image guidance performed on the 2D intraoperative X-rays.

Furthermore, we take advantage of using a differentiable DRR rendering process [11] to complement this geometric loss with an intensity-based loss that compares the input DRR with the synthetic DRRs simulated using the initial and final estimated poses. Inspired by previous works, we adopt here a multiscale normalized cross-correlation (mNCC) loss $\mathcal{L}_{\mathrm{mNCC}}$, due to its effectiveness in capturing both structural and intensity correspondences [10].

In summary, LXPose is trained end-to-end using the following loss function:

$$\mathcal{L} = (\lambda \mathcal{L}_{\mathrm{mPD}} + \mathcal{L}_{\mathrm{mNCC}})_1 + (\lambda \mathcal{L}_{\mathrm{mPD}} + \mathcal{L}_{\mathrm{mNCC}})_2, \qquad (2)$$

where the first and second terms are computed on the outputs of CNN1 and CNN2, respectively, and λ balances the effect of $\mathcal{L}_{\mathrm{mPD}}$ and $\mathcal{L}_{\mathrm{mNCC}}$.

2.5 DRR/X-Ray Domain Adaptation with Data Augmentation

To address the domain shift between the training DRRs and the real X-rays seen at test-time, we apply aggressive data augmentation: Gaussian noise, Gaussian blur, sharpening, and gamma transforms. In addition to these augmentations commonly used in the CT/X-ray registration literature, we introduce two new X-rayspecific augmentations to further reduce the domain gap. First, we include salt-and-pepper noise (SP) to simulate defective sensors in the X-ray detector. Then, we leverage plasma intensity variations [28] to render inhomogeneities in X-ray contrasts due to beam hardening and the anode heel effect [5]. More precisely, plasma augmentations use the diamond-square algorithm [7] to generate fractal maps with multi-scale, spatially varying intensities. These maps are then applied to the training DRRs with element-wise multiplication.

2.6 Implementation Details

DRR Simulation. Training DRRs are simulated on the fly using DiffDRR [11]. Here, we place the source at the origin and the detector at $(0, 0, f)$. We reorient the depth axis to align with the CT's posterior-anterior axis, and translate the source and detector along the z-axis by 750 mm to place them around the subject. DRRs are then simulated by sampling C-arm pose parameters from normal centered distributions with the following standard deviations: 15° (α, β), 5° (γ), 50 mm (t_x, t_y), 100 mm (t_z). As in previous works [9,37], we simulate DRRs at reduced size (256×256) for memory usage.

Architecture and Training. Each CNN consists in a ResNet-18 [17] followed by a fully connected layer. Based on validation scores, we set $\lambda = 10^{-1}$. Training is conducted for 2000 epochs with a batch size of 16 using Adam [20] with an initial learning rate of 10^{-3} and a warmup cosine scheduler [35]. Our network is implemented using PyTorch [29]. Training takes 20 h on a 80Gb Nvidia A100 GPU, and inference runs in 20 ms on a 12Gb Nvidia 1080Ti GPU.

3 Experiments and Results

3.1 Dataset and Preprocessing

We evaluate our pipeline on the DeepFluoro dataset [14], an open-source collection of six cadaveric subjects, each comprising a CT volume as well as an X-ray series between 24 and 111 images (366 in total). The C-arm is fully calibrated with known intrinsic parameters, and every X-ray is provided with ground truth extrinsic camera parameters, only used for testing purposes. Moreover, CTs are available with 14 manual anatomical 3D landmarks and the coordinates of their corresponding 2D projections in the test X-rays. Real X-rays are held-out for testing, whereas training and validation are performed on patient-specific synthetic DRRs to relax supervision requirements and to simulate clinical settings where networks are trained preoperatevely without available X-rays.

Here, we adopt the X-ray absorption convention to match the DiffDRR formulation. Hence, we convert raw intraoperative attenuation images I into absorption images $I_\mu = \log[\max(I)] - \log I$, where the brightest pixels $\max(I)$ correspond to rays that travel through air only. Finally, we crop all images to remove the collimator effect, and resample them to 256×256.

3.2 Competing Methods

The proposed LXPose approach is compared against 3 baseline methods:
PoseNet is a network trained on real X-rays to regress C-arm poses from ground truth poses [19]. Since the associated code is not published, we re-implement PoseNet and train it on real X-rays using leave-one-subject-out cross-validation.
xvr-regressor is the regression CNN used in xvr [9], and differs from LXPose by: not being multi-stage, replacing $\mathcal{L}_{mPD}$ by a geodesic loss, regressing Euler angles instead of axis-angles, and not using the newly introduced augmentations.
xvr is the state-of-the-art method proposed by [9] with regression and optimization steps. We test it for reference only as it far exceeds the 40 ms time limit.

For fair comparison, all networks are trained using the same architecture (i.e., ResNet-18), preprocessing, and generic augmentations as LXPose.

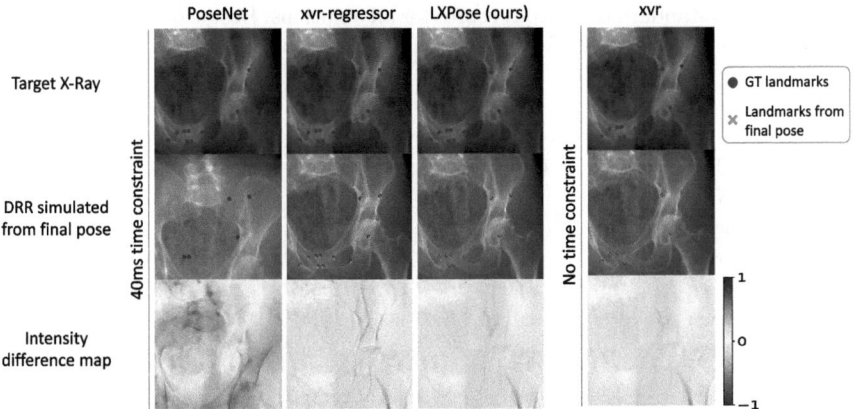

Fig. 2. CT/X-ray registration results for all methods visualized as: the target X-ray with 3D-derived landmarks projected from the ground truth pose, DRR with the landmarks projected from the predicted pose, and intensity differences between the two.

Table 1. Mean (standard deviation) scores for all methods. Best scores are in bold and stars mark statistical significance (Bonferroni-corrected Wilcoxon signed-rank test, 1% level). All differences between LXPose and xvr are also significant (same test).

Metric		PoseNet [19]	xvr-regressor [9]	LXPose (ours)		xvr [9]
mPD (mm)		77.70 (22.58)	4.07 (2.06)	**1.25 (1.04)***		0.62 (1.24)
mTRE (mm)		187.18 (22.58)	32.41 (18.29)	**13.52 (10.48)***		4.74 (11.39)
α (˘r)	40ms time constraint	10.42 (7.31)	1.35 (1.18)	**0.46 (0.46)***	No time constraint	0.21 (0.45)
β (˘r)		5.90 (4.12)	1.55 (1.30)	**0.73 (0.78)***		0.25 (0.75)
γ (˘r)		7.39 (3.30)	0.98 (0.73)	**0.33 (0.35)***		0.08 (0.16)
t_x (mm)		15.90 (11.81)	1.20 (0.93)	**0.33 (0.37)***		0.14 (0.24)
t_y (mm)		48.00 (16.37)	1.03 (0.78)	**0.33 (0.27)***		0.14 (0.31)
t_z (mm)		62.91 (37.56)	12.67 (9.56)	**4.73 (5.07)***		1.86 (3.73)
time (s)		0.007 (0.032)	0.007 (0.032)	0.020 (0.033)		7.777 (2.094)

3.3 Results

Registration Performances. We evaluate our results using mean distances between the registered and ground truth landmarks, both in 3D with mTRE [22] and in 2D (i.e., after projection) with mPD. Note that mTRE is reported for completeness since mPD is more insightful for image guidance in the 2D intraoperative X-ray space. We also assess mean absolute errors of the estimated pose parameters $\alpha, \beta, \gamma, t_x, t_y, t_z$. Finally, we report the inference time per image, which is critical for real-time image guidance.

While xvr yields the best scores for all metrics (Table 1), inference requires more than 7 s due to the optimization step, making it unsuitable for real-time use. In comparison, LXPose offers a favorable trade-off: although xvr obtains slightly better results (difference of 0.6 mm in mPD), LXPose remains highly accurate (see Fig. 2) while yielding a mean inference time of just 20 ms, thus complying

Table 2. Mean (standard deviation) scores for all ablations. Best scores are in bold and * marks stat. significance (Bonferroni-corrected Wilcoxon signed-rank test, 1% level).

Ablations	real test X-rays		synthetic test DRRs	
	mPD (mm)	mTRE (mm)	mPD (mm)	mTRE (mm)
LXPose (ours)	**1.25 (1.04)***	13.51 (10.47)	**0.55 (0.66)***	**3.02 (2.20)***
LXPose ⊖ CNN2	1.75 (1.39)	17.81 (13.83)	1.33 (1.54)	7.80 (5.38)
LXPose, sequential training	1.75 (1.35)	17.51 (13.35)	1.45 (1.56)	7.91 (4.90)
LXPose, Euler angles	2.48 (1.85)	19.72 (17.08)	1.99 (2.38)	11.8 (10.00)
LXPose, geodesic loss	2.60 (3.60)	**11.73 (10.82)***	1.82 (1.65)	3.08 (2.36)
LXPose ⊖ plasma and SP aug.	4.01 (1.82)	45.61 (19.75)	0.75 (0.65)	5.14 (2.96)
LXPose ⊖ all (= xvr-regressor)	4.07 (2.06)	32.41 (18.29)	2.19 (2.76)	9.43 (12.31)

with the 40 ms time constraint for intraoperative deployment. Interestingly, the gap in mTRE is substantially larger than mPD. This discrepancy is mainly driven by errors along the 3D depth axis parameter t_z, which primarily affect magnification but have a limited impact once projected onto the 2D image plane. Finally, LXPose significantly outperforms every baseline that satisfy the 40 ms time constraint, with, for example, gaps of 2.82 mm and 76.45 mm in mPD for xvr-regressor and PoseNet, respectively. These results highlight the quality of our predictions in the allowed time budget for real-time guidance.

Ablation Study. We now conduct ablations to study the impact of each component of LXPose (Table 2). First, removing the network CNN2 leads to a significant performance drop, thus highlighting the role of our multi-stage formulation for progressive refinement. Then, we observe that training end-to-end improves accuracy by enabling CNN1 and CNN2 to learn synergies, as opposed to sequential training. Further, regressing axis-angle representation of r yields more robust estimates than using Euler angles. Finally, while the geodesic loss employed in [9] yields narrowly better mTRE scores (gap of 1.7 mm), our loss shows better mPD scores (1.3 mm gap), which is the metric of choice for 2D image guidance.

Domain Gap Quantification. In this last experiment, we precisely quantify the domain gap between the target X-ray domain and the training DRR domain. To this end, we leverage the ground truth C-arm pose associated with each real X-ray to simulate a corresponding test DRR. As expected, all networks perform better on the DRR source domain (0.92 mm mPD difference with real X-rays across methods). Interestingly, ablating the newly introduced plasma and SP augmentations leads to a far greater gap (3.26 mm). This ablation also mainly explains our improvements over xvr-regressor (i.e., fully ablated LXPose), thus showing the crucial role of our augmentations in tackling the domain gap.

4 Conclusion

We have presented LXPose, an end-to-end deep learning framework for real-time 3D/2D registration of preoperative CTs to intraoperative X-rays. Remarkably, LXPose almost reaches the projection accuracy of the state-of-the-art method, but runs considerably faster to enable real-time image-guidance. This has been achieved by leveraging deep learning for speed, a multi-stage CNN for progressive pose estimation, a 2D loss for accurate projection in the intraoperative image space, and an aggressive augmentation pipeline for enhanced robustness on real data. Future work will first assess our method on additional datasets (e.g., interventional cardiology), possibly with real-time deployment. To this end, we will also aim to further bridge the DRR/X-ray domain gap to improve robustness on real X-rays. Finally, we will extend our method to incorporate pre-training on multi-patient data to increase generalization [9]. By providing precise real-time guidance, LXPose has the potential to reduce procedure durations, alleviate the cognitive burden of clinicians, and improve interventional outcomes.

Acknowledgments. This work has been supported by the French government, through the 3IA Cote d'Azur Investments in the project managed by the National Research Agency (ANR) with the reference number ANR-23-IACL-0001. Further support was provided with computer resources by GENCI at CINES/IDRIS thanks to the grant 2025- A0180315107 on the supercomputer Jean Zay's V100/A100 Partition.

Disclosure of Interests. The authors have no competing interests to declare.

References

1. Berger, M., et al.: Marker-free motion correction in weight-bearing cone-beam CT of the knee joint. Med. Phys. **43**(3), 1235–1248 (2016)
2. Bier, B., et al.: X-ray-transform invariant anatomical landmark detection for pelvic trauma surgery. In: Frangi, A.F., Schnabel, J.A., Davatzikos, C., Alberola-López, C., Fichtinger, G. (eds.) MICCAI 2018. LNCS, vol. 11073, pp. 55–63. Springer, Cham (2018). https://doi.org/10.1007/978-3-030-00937-3_7
3. Bui, M., Albarqouni, S., Schrapp, M., Navab, N., Ilic, S.: X-Ray PoseNet: 6 DoF pose estimation for mobile X-ray devices. In: WACV, pp. 1036–1044 (2017)
4. Chen, J., et al.: A survey on deep learning in medical image registration: New technologies, uncertainty, evaluation metrics, and beyond. Med. Image Anal. **100** (2025)
5. Chou, M.C.: Evaluation of non-uniform image quality caused by anode heel effect in digital radiography using mutual information. Entropy **23**, 525 (2021)
6. Esteban, J., Grimm, M., Unberath, M., Zahnd, G., Navab, N.: Towards fully automatic X-Ray to CT registration. In: Shen, D., et al. (eds.) MICCAI 2019. LNCS, vol. 11769, pp. 631–639. Springer, Cham (2019). https://doi.org/10.1007/978-3-030-32226-7_70
7. Fournier, A., Fussell, D., Carpenter, L.: Computer rendering of stochastic models. Commun. ACM **25**(6), 371–384 (1982)
8. Geng, H., et al.: CT2X-IRA: CT to X-ray image registration agent using domain-cross multi-scale-stride deep reinforcement learning. Phys. Med. Biol. **68**(17) (2023)

9. Gopalakrishnan, V., et al.: Rapid patient-specific neural networks for intraoperative X-ray to volume registration. arXiv:2503.16309 (2025)
10. Gopalakrishnan, V., Dey, N., Golland, P.: Intraoperative 2D/3D image registration via differentiable X-ray rendering. In: CVPR, pp. 11662–11672 (2024)
11. Gopalakrishnan, V., Golland, P.: Fast auto-differentiable digitally reconstructed radiographs for solving inverse problems in intraoperative imaging. In: Workshop on Clinical Image-Based Procedures (2022)
12. Grimm, M., Esteban, J., Unberath, M., Navab, N.: Pose-dependent weights and domain randomization for fully automatic X-ray to CT registration. IEEE Trans. Med. Imaging **40**(9), 2221–2232 (2021)
13. Groher, M.: 2D-3D registration of vascular images: towards 3D-guided catheter interventions. Ph.D. thesis, Technische Universität München (2008)
14. Grupp, R., et al.: Automatic annotation of hip anatomy in fluoroscopy for robust and efficient 2D/3D registration. Int. J. Comput. Assist. Radiol. Surg. **15**(5), 759–769 (2020)
15. Gu, W., Gao, C., Grupp, R., Fotouhi, J., Unberath, M.: Extended capture range of rigid 2D/3D registration by estimating riemannian pose gradients. In: International Workshop on Machine Learning in Medical Imaging, pp. 281–291 (2020)
16. Hartley, R.: Multiple View Geometry in Computer vision, vol. 665. Cambridge University Press (2003)
17. He, K., Zhang, X., Ren, S., Sun, J.: Deep residual learning for image recognition. In: CVPR, pp. 770–778 (2016)
18. Jaganathan, S., Kukla, M., Wang, J., Shetty, K., Maier, A.: Self-supervised 2D/3D registration for X-ray to CT image fusion. In: WACV, pp. 2788–2798 (2023)
19. Kendall, A., Grimes, M., Cipolla, R.: PoseNet: a convolutional network for real-time 6-DOF camera relocalization. In: ICCV, pp. 2938–2946 (2015)
20. Kingma, D., Ba, J.: Adam: a method for stochastic optimization. arXiv:1412.6980 (2014)
21. Knaan, D., Joskowicz, L.: Effective intensity-based 2D/3D rigid registration between fluoroscopic X-ray and CT. In: Ellis, R.E., Peters, T.M. (eds.) MICCAI 2003. LNCS, vol. 2878, pp. 351–358. Springer, Heidelberg (2003). https://doi.org/10.1007/978-3-540-39899-8_44
22. van de Kraats, E.B., Penney, G.P., Tomaževič, D., van Walsum, T., Niessen, W.J.: Standardized evaluation of 2D-3D registration. In: Barillot, C., Haynor, D.R., Hellier, P. (eds.) MICCAI 2004. LNCS, vol. 3216, pp. 574–581. Springer, Heidelberg (2004). https://doi.org/10.1007/978-3-540-30135-6_70
23. Lee, B., et al.: Breathing-compensated neural networks for real time C-arm pose estimation in lung CT-fluoroscopy registration. In: ISBI (2022)
24. Lemieux, L., Jagoe, R., Fish, D., Kitchen, N., Thomas, D.: A patient-to-computed-tomography image registration method based on digitally reconstructed radiographs. Med. Phys. **21**, 1749–1760 (1994)
25. Meng, C., Wang, Q., Guan, S., Sun, K., Liu, B.: 2D–3D registration with weighted local mutual information in vascular interventions. IEEE Access **7**, 629–638 (2019)
26. Miao, S., Wang, Z.J., Liao, R.: A CNN regression approach for real-time 2D/3D registration. IEEE Trans. Med. Imaging **35**(5), 1352–1363 (2016)
27. Nguyen, V., et al.: Automatic landmark detection and mapping for 2D/3D registration with BoneNet. Front. Veterinary Sci. **9**, 923449 (2022)
28. Nicolaou, A., Christlein, V., Riba, E., Shi, J., Vogeler, G., Seuret, M.: TorMentor: deterministic dynamic-path, data augmentations with fractals (2022)
29. Paszke, A., et al.: Pytorch: an imperative style, high-performance deep learning library. In: NeurIPS, vol. 32 (2019)

30. Pfandler, M., Stefan, P., Mehren, C., Lazarovici, M., Weigl, M.: Technical and nontechnical skills in surgery: a simulated operating room environment study. Spine **44**(23), 1396–1400 (2019)
31. Shrestha, P., Xie, C., Shishido, H., Yoshii, Y., Kitahara, I.: X-ray to CT rigid registration using scene coordinate regression. Greenspan, H., et al. (eds.) MICCAI 2023. LNCS, vol. 14229, pp. 781–790. Springer, Cham (2023). https://doi.org/10.1007/978-3-031-43999-5_74
32. Unberath, M., et al.: The impact of machine learning on 2D/3D registration for image-guided interventions: a systematic review and perspective. Front. Robot. AI **8** (2021)
33. Unberath, M., et al.: Enabling machine learning in X-ray-based procedures via realistic simulation of image formation. Int. J. Comput. Assist. Radiol. Surg. **14**, 1517–1528 (2019)
34. Wang, J., et al.: Dynamic 2-D/3-D rigid registration framework using point-to-plane correspondence model. IEEE Trans. Med. Imaging **36**(9), 1939–1954 (2017)
35. Wolf, T., et al.: HuggingFace's Transformers: state-of-the-art natural language processing. arXiv:1910.03771 (2019)
36. Zaffino, P., Moccia, S., De Momi, E., Spadea, M.: A review on advances in intraoperative imaging for surgery and therapy: imagining the operating room of the future. Ann. Biomed. Eng. **48**(8), 2171–2191 (2020)
37. Zhang, B., Faghihroohi, S., Azampour, M., Liu, S., Navab, N.: A patient-specific self-supervised model for automatic X-Ray/CT registration. In: Greenspan, H., et al. (eds.) MICCAI 2023. LNCS, vol. 14228, pp. 515–524. Springer, Cham (2023). https://doi.org/10.1007/978-3-031-43996-4_49
38. Zheng, J., Miao, S., Wang, J., Liao, R.: Pairwise domain adaptation module for CNN-based 2-D/3-D registration. J. Med. Imaging **5**(2) (2018)

X-RAFT: Cross-Modal Non-rigid Registration of Blue and White Light Neurosurgical Hyperspectral Images

Charlie Budd[1]([✉]), Silvère Ségaud[1], Matthew Elliot[1,2], Graeme Stasiuk[3], Yijing Xie[1], Jonathan Shapey[1,2], and Tom Vercauteren[1]

[1] Department of Surgical & Interventional Engineering, School of Biomedical Engineering & Imaging Sciences, King's College London, London SE1 7EH, UK
charles.budd@kcl.ac.uk
[2] Department of Neurosurgery, King's College London Hospital NHS Foundation Trust, London SE5 9RS, UK
[3] Department of Imaging Chemistry & Biology, School of Biomedical Engineering & Imaging Sciences, King's College London, London SE1 7EH, UK

Abstract. Integration of hyperspectral imaging into fluorescence-guided neurosurgery has the potential to improve surgical decision making by providing quantitative fluorescence measurements in real-time. Quantitative fluorescence requires paired spectral data in fluorescence (blue light) and reflectance (white light) mode. Blue and white image acquisition needs to be performed sequentially in a potentially dynamic surgical environment. A key component to the fluorescence quantification process is therefore the ability to find dense cross-modal image correspondences between two hyperspectral images taken under these drastically different lighting conditions. We address this challenge with the introduction of X-RAFT, a Recurrent All-Pairs Field Transforms (RAFT) optical flow model modified for cross-modal inputs. We propose using distinct image encoders for each modality pair, and fine-tune these in a self-supervised manner using flow-cycle-consistency on our neurosurgical hyperspectral data. We show an error reduction of 36.6% across our evaluation metrics when comparing to a naive baseline and 27.83% reduction compared to an existing cross-modal optical flow method (Cross-RAFT). Our code and models are publicly available (https://github.com/charliebudd/x-raft-cross-modal-non-rigid-registration).

Keywords: Quantitative Fluorescence · Hyperspectral Imaging · Image Registration · Optical Flow · Cross Modality

1 Introduction

Fluorescence guided surgery (FGS) is often employed in situations where tissue differentiation is both difficult and critical to patient outcomes. The efficacy of FGS for glioma surgery has led to its wide adoption across neurosurgery units [3].

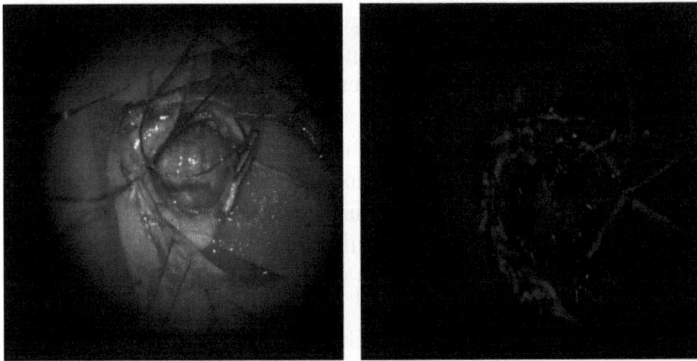

Fig. 1. Exemplar white-blue image pair from our neurosurgical dataset showing a high-grade glioma visible through a craniotomy surrounded by surgical patties. Both images have been converted to RGB images for visualisation. The blue light image reveals a high level of pink fluorescence from the glioma, as-well as naturally occurring green auto-fluorescence from other tissues. This image pair demonstrates both homographic motion due to movement of the surgical microscope and minor non-homographic motion due to the motion of the surgical patty threads. (Color figure online)

A fluorescence-inducing drug (5-ALA) is administered prior to surgery, leading to a build-up of fluorophore (PpIX) in tumour tissue which then fluoresces pink under blue light [10]. Neurosurgical 5-ALA-PpIX FGS is not without limitations as surgeons still make qualitative judgment based on visual inspection of the margins of fluorescence glow, as illustrated in Fig. 1. Integration of hyperspectral imaging (HSI) into 5-ALA-PpIX FGS has the potential to provide real-time quantitative fluorescence measurements at every pixel across the field of view [6]. However, accurate fluorescence quantification in turbid biological tissue requires compensating for the varying optical characteristics of tissue [12]. Combining paired spectral data from blue light fluorescence and white-light reflectance has been shown to allow for such compensation [13]. Due to the dynamic nature of surgery combined with the fact that white light reflectance and blue light PpIX fluorescence HSI cannot be acquired simultaneously, it cannot be guaranteed that any pair of white and blue light HSI data is spatially well aligned. The white light image must therefore be spatially co-registered with the blue light image, requiring dense pixel-wise spatial correspondences. In other words, cross-modal optical flow is needed to achieve quantitative FGS in neurosurgery.

Deep learning methods dominate the field of optical flow estimation. Recurrent All-Pairs Field Transforms (RAFT) [11] represents the state-of-the-art approach for it and it demonstrates solid generalisability to surgical images [1,4,9]. Optical flow methods rely on local similarity between corresponding image regions. This presents a challenge when working with cross-modal image pairs, as the image regions to be matched cannot trivially be compared. While some recent works have focused on cross-modal homography estimation [5,7], this is not sufficient as our image pairs feature a significant amount of non-homographic motion. Application of multimodal registration designed for medical imaging data [2] doesn't translate easily to surgical images due to large tissue motion and the projective nature of surgical microcopy views. Closer to our task, CrossRAFT [15] and CrossMCMRAFT [14], both attempt to train optical flow models to be robust to an open-set of modalities by training with single modality image pairs with differing random augmentations applied. This approach is attractive for a number of reasons, namely that image pairs from a single modality may be used for training, and that the resulting model would in theory generalise to any pair of modalities. However, this generalisability is heavily dependant on the augmentations employed.

In this work, we propose X-RAFT, the first learning-based cross-modal optical flow approach in a surgical setting. Our main novel contribution, lies in leveraging distinct modality-specific image encoders within a modified RAFT architecture. This is in contrast to CrossRAFT and CrossMCMRAFT which both take the approach of attempting to unify the encoded latent representations. Another main contribution, key to enabling our objectives in the abscence of large-scale ground-truth data, is the introduction of a self-supervised training approach that requires only unaligned image triplets of mixed modality. Our training approach allows X-RAFT to learn intricacies of the modalities beyond what domain randomisation can achieve. Our experiments show that X-RAFT provides significant performance increase over our baselines.

2 Material and Methods

Neurosurgical Dataset. Our data consists of 52 hyperspectral videos of glioma surgeries. The videos were captured with a 10 band hyperspectral camera integrated into a surgical microscope as part of an ethically-approved clinical study[1]. Each video consist of sections recorded under white light illumination followed by blue light illumination used by the surgeon for fluorescence visualisation. Due to the low quantum efficiency of PpIX, exposure times in the range of 1.5 to 2 s are used to capture blue light HSI. Even then, the blue images present a lower signal-to-noise-ratio than the white counterparts, as shown in Fig. 1 where colour imaging is reconstructed from high-dimensional HSI data [8].

While surgeons were instructed to target static scenes during the recording, the reality of recording in a high-stakes environment meant that around 60% of the data contains a noticeable level of motion. This motion arises from a number of factors, including handling of surgical tools, manipulation of brain tissue, pulsatility of the patient brain, and sometimes repositioning of the surgical microscope. Through visual inspection, we manually split our dataset into 32 videos with clearly visible motion and 20 videos with little-to-no motion. We manually selected around 10–20 white and blue images per video based on various qualitative assessments such as lack of motion blur and if the image is well exposed. We then annotated our high motion data to aid in evaluation. Specifically, a trained neurosurgeon selected a white+blue image pair from each

---

[1] Ethics information redacted for the anonymous review process.

high motion video, forming a reserved testing split. They then created binary segmentations of visible tumour tissue and marked several corresponding key point pairs of various elements in both the white and the blue frame.

Baseline Approaches. We choose RAFT as a baseline due to its availability, popularity, and generalisable performance to surgical scenarios. As RAFT is intended for RGB images, not hyperspectral data, we first convert our white light and blue light HSI data to RGB images by applying a colour space conversion matrix. Different colour spaces can be targeted for different purposes, with standard RGB (sRGB) allowing intuitive human vision like visualisation [8]. A naive approach to bridge the domain gap between our blue and white images is to apply different handcrafted colour transformations to each image to increase the pixel-wise similarity of the RGB reconstruction. There are two main contributors of signal in our blue light HSI data: pink and green fluorescence and residual blue reflectance not rejected by the excitation light rejection filter. Conversely, our white light HSI contain only reflectance data but from across the visible spectral range. As such, only blue reflectance data is common signal across the two modalities. We therefore choose to take the blue channel from both reconstructed sRGB images, repeating this three times to construct greyscale (BBB) images for inputting into RAFT. We present results for RAFT running on both sRGB and BBB images in the results section. We additionally make use of CrossRAFT [15] as a additional baseline specifically aimed at cross-modal inference. We tried both the pretrained weights provided as well as running their published training pipeline on our neurosurgical data. The former produced the best results in all metrics by a fair margin and so we report only these results for brevity. Unfortunately CrossMCMRAFT [14] cannot be compared as there is no implementation or model released with this work.

Our Proposed X-RAFT. RAFT estimates optical flow through a three-stage process. Firstly, the source and target images are independently passed through a feature encoder to produce latent feature embeddings with reduced spatial resolution. These embeddings are then used to construct a dense all-pairs correlation volume, capturing potential correspondences between every location in the two images. Finally, the optical flow map is generated through a series of iterative updates guided by a separate encoding of the source image, produced by a context encoder. Due to the significant difference in local image features between our white and blue images, we hypothesise that it will be beneficial to learn different feature and context encoders for these modalities. We extend this thinking further by learning a distinct feature and context encoder for each possible modality pair (e.g. blue-to-blue, blue-to-white, white-to-blue, and white-to-white). In this way, each encoder can focus on extracting features that will best help find correspondences in the other modality. As an example, to infer optical flow from a white source image to a blue target image, as depicted in Fig. 2, the white source image is encoded by the white-to-blue feature and context encoders, while the blue target image is encoded by the blue-to-white feature encoder. Note that

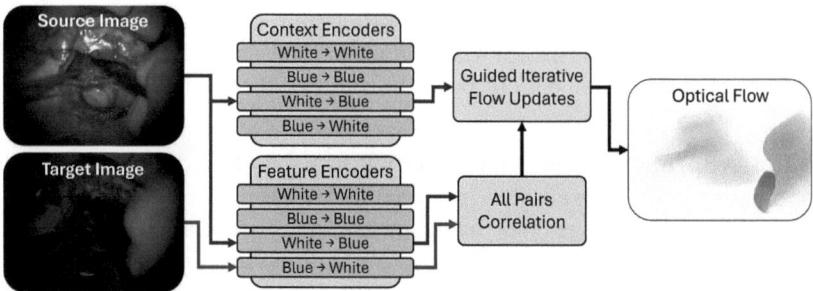

Fig. 2. Model architecture of X-RAFT. Our modifications to the original RAFT architecture simply add duplicate feature and context encoders for each modality pair.

we do not make this directional, i.e. the white-to-blue encoder is used to encode white images whether they are the source or the target image, so long as the other image in the pair is a blue image.

Scenario-Specific Choices. For our naive baseline method, we reasoned that the blue channel is of most importance to finding correspondences. We would like to provide this prior information to the model whilst still allowing the model to learn helpful features which may be extracted from the other channels. To achieve this, we carefully initialise the weights of the first convolutional layer for all cross-modal encoders to produce the same output for an RGB image as if we had instead provided a BBB image. In the resulting X-RAFT RGB, the weights B are zeroed for the red and green channels, while the blue channel is assigned the channel-wise sum of the pre-trained weights A:

$$B_{niwh} = \begin{cases} \sum_i A_{niwh}, & \text{if } i = 3 \\ \mathbf{0}, & \text{otherwise} \end{cases} \quad (1)$$

As our 10 band HSI data contains information beyond our RGB reconstruction, we also propose X-RAFT HSI which accepts 10 input channels. We again modify the first layer of each image encoder. The weights of this layer C are efficiently initialised by multiplying the existing weights B by the colour space conversion matrix Q, used to convert the hyperspectral images into RGB images.

$$C_{ncwh} = \sum_i B_{niwh} \cdot Q_{ic} \quad (2)$$

This provides the same results as running on RGB images but removes the 3 channel bottleneck, thus allowing the model to learn to use the new channel information without degrading initial performance.

Flow-Cycle-Consistency Loss. As we lack ground truth optical flow for our data, we require a self-supervised training strategy. Typical methods in optical flow

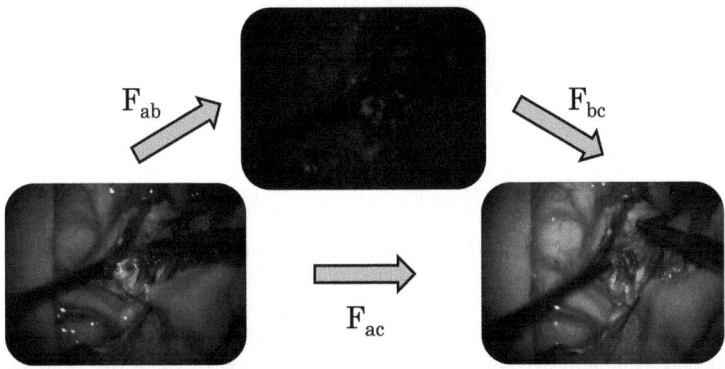

Fig. 3. Depiction of the flow cycle-consistency triplet for our self-supervision X-RAFT training. Optical flow inferred between the two white images, using off-the-shelf RAFT, is used as supervision for training of the white-to-blue and blue-to-white flow. (Color figure online)

rely on either flow consistency or photometric constancy. In our case, photometric constancy is inappropriate as the image pairs would not be expected to be visually similar after registration. As such, we choose to design a flow consistency based learning objective. We take advantage of the fact that we posses multiple white light images of each scene, between which we may infer reliable optical flow using off-the-shelf pre-trained RAFT. We therefore construct a flow cycle around a triplet constituting of two different white images and a blue image sampled from the same video, depicted in Fig. 3. Our loss function may then be expressed as the Euclidean distance between the vectors in the two optical flow fields, known as the end point error (EPE):

$$L = \text{EPE}(F_{ab} + \text{warp}(F_{bc}, F_{ab}), F_{ac}) \tag{3}$$

where F_{ij} represents the optical flow from the image I_i to the image I_j, and warp(A, F) pulls pixels in the entity A backwards along the flow field F through grid sampling. Spatial correspondences between two images may not exist everywhere or information may be missing to extract them meaningfully. Our training objective is thus adapted by attempting to mask out such regions. One main consideration is occlusion induced by motion between frames. A typical approach for determining occluded regions is to calculate the optical flow in both directions. If true correspondences are found, the flow fields should have low discrepancy:

$$M_{ij}^o = (F_{ij} - \text{warp}(F_{ji}, F_{ij})) > \varepsilon^o \tag{4}$$

Additionally we note that due to the low signal in much of the blue image, a significant part of the image can be within the noise floor (i.e. black). We define a second mask by thresholding the average pixel intensity across all channels:

$$M_i^d = \frac{1}{C} \sum_{c=1}^{C} I_{i,c} > \varepsilon^d \tag{5}$$

For consistency, the channel average is computed from HSI, even if the image is then converted into RGB for inputting to the model. This determines where there is little to no feature detail to use in finding correspondences. A final mask relative to I_a may then be calculated as the combination (pixel-wise multiplication) of these masks, warped where appropriate to align with the source image:

$$M = M_{ac}^o \cdot M_{ab}^o \cdot \text{warp}(M_{bc}^o M_b^d, F_{ab}). \tag{6}$$

This mask is then used to mask our loss function so that we only provide supervision for image regions we believe to have findable correspondences.

Model Training. To avoid overfitting to our approximate training objective and small dataset, we choose to freeze all parameters of the model, leaving only the cross-modal feature and context encoders unfrozen. We iterate through our high motion training data sampling batches of white-blue-white triplets and optimising our flow-cycle-consistency loss for the first three flow field iterations. Mask thresholds of 8.0 and 0.07 were used for ε^o and ε^d respectively. We used a batch size of 20 and an Adam optimiser with a learning rate of 5×10^{-5}. We validate against our synthetic flow metric, as described in the Evaluation Section, every 10 batches and early stop with a patience of 20, generally our models trained for around 450 batches taking about an hour on a single 40 Gb A100 GPU. All hyper parameters were decided on via a light and informally hyper parameter tuning based on the validation performance.

Evaluation. Due to the lack of ground truth pixel-wise correspondences, we choose to employ a mixed evaluation strategy using approximate, sparse, or secondary metrics to ensure robust method ranking. Each metric is evaluated in both inference directions, white to blue and blue to white, as well as the combination of the two. Firstly, we generate synthetic flow fields which are then used to apply elastic deformation to one image from a white-blue pair sampled from our low motion dataset. We make the assumption that the only motion between the image pair is that which is induced by the deformation, thus the generated optical flow field provides pseudo ground truth. Following RAFT, we use endpoint error (EPE) to measure the accuracy for the predicted optical flow. While this provides dense error measurement, the generated motion is not necessarily realistic and may not account for all motion between the images. To address this we also make use of our annotated image pairs from our high motion dataset. The annotated key point pairs are used to provide sparse but precise evaluation of the inferred flow fields, again using EPE. Finally, the inferred optical flow fields are then used to propagate the annotated tumour segmentation masks between the two images. We take 1−IoU of the annotated and propagated mask as an error metric for consistency with the other error metrics.

3 Results and Discussion

We present evaluation metrics for all methods in Table 1. This shows that X-RAFT running on hyperspectral images provides an average error reduction across the three metrics of 36.6% compared to RAFT on our transformed BBB images and 27.8% compared to CrossRAFT on RGB images. Discarding the synthetic flow result from the calculation of percentage improvements, for which CrossRAFT performs uncharacteristically poorly, we still find a 9.0% improvement. In addition we present an exemplar blue-white registration in Fig. 4 demonstrating an increase in found correspondences during registration.

Table 1. Evaluation metrics for different optical flow models presented as means and standard deviations over 5 training runs where appropriate. Scores are reported separately for white-to-blue and blue-to-white inference directions as well as the average.

Model	Direction	Synthetic (EPE ↓)	Keypoint (EPE ↓)	Mask (1−IoU ↓)
RAFT RGB	W→B	26.636	6.050	0.260
	B→W	19.617	12.980	0.373
	Both	23.127	9.515	0.316
RAFT BBB	W→B	8.084	6.149	0.230
	B→W	6.815	10.972	0.364
	Both	7.449	8.561	0.297
CrossRAFT RGB	W→B	11.475	6.411	0.200
	B→W	16.323	6.241	0.186
	Both	13.899	6.313	0.193
X-RAFT RGB	W→B	5.183 ± .107	5.649 ± .226	**0.181 ± .004**
	B→W	4.484 ± .021	5.719 ± .073	**0.184 ± .005**
	Both	4.834 ± .046	5.684 ± .137	**0.182 ± .004**
X-RAFT HSI	W→B	**5.112 ± .094**	**5.275 ± .229**	0.184 ± .004
	B→W	**4.466 ± .023**	**5.608 ± .115**	0.185 ± .002
	Both	**4.789 ± .047**	**5.442 ± .129**	0.185 ± .003

In this work we have identified a novel integration problem requiring the first investigation of learning-based cross-modal optical flow in a surgical setting. The novel architectural modifications and self-supervised fine-tuning methodology of our proposed X-RAFT model demonstrates a clear improvement over the existing state-of-the-art approach. A primary limitation of this work is the use of frames from the same videos as frames used to train the models. This was done to ensure a significant amount of evaluation data was available. While we aimed to prevent overfitting by validating on the low-motion data not used in training, and by only training the encoders, it is still possible that the high correlation with the training data may have introduced bias. Our future work will see the application of this methodology to aid in the quantification of fluorescence measurement by allowing the use of the co-register white light images to compensate for optical properties in the blue light image space.

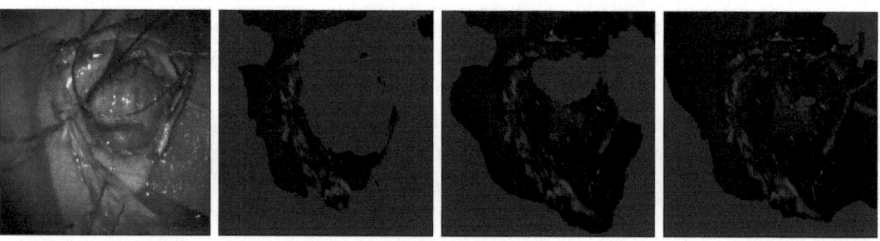

Fig. 4. An example registration of a blue image onto a white image using, from left to right, RAFT BBB, CrossRAFT, and X-RAFT HSI. Flow discrepancy with a threshold of 3 pixels is used to discard uncertain correspondences (shown in gray). (Color figure online)

Acknowledgments. This work was supported by funding from NIHR [NIHR202114], and Wellcome [WT223880/Z/21/Z]. For the purpose of open access, the authors have applied a CC BY public copyright licence to any Author Accepted Manuscript version arising from this submission.

Disclosure of Interests. TV and JS are co-founders and shareholders of Hypervision Surgical.

References

1. Budd, C., Vercauteren, T.: Transferring relative monocular depth to surgical vision with temporal consistency. In: International Conference on Medical Image Computing and Computer-Assisted Intervention, pp. 692–702. Springer (2024)
2. Chen, J., et al.: A survey on deep learning in medical image registration: new technologies, uncertainty, evaluation metrics, and beyond. Med. Image Anal. **100**, 103385 (2025)
3. Elliot, M., et al.: Fluorescence guidance in glioma surgery: a narrative review of current evidence and the drive towards objective margin differentiation. Cancers **17**(12), 2019 (2025)
4. Gerats, B.G.A., Wolterink, J.M., Mol, S.P., Broeders, I.A.M.J.: Neural fields for 3d tracking of anatomy and surgical instruments in monocular laparoscopic video clips (2024). https://arxiv.org/abs/2403.19265
5. Jiang, Z., Zhang, Z., Liu, J., Fan, X., Liu, R.: Breaking modality disparity: harmonized representation for infrared and visible image registration. arXiv preprint arXiv:2304.05646 (2023)
6. Kotwal, A., Saragadam, V., Bernstock, J.D., Sandoval, A., Veeraraghavan, A., Valdés, P.A.: Hyperspectral imaging in neurosurgery: a review of systems, computational methods, and clinical applications. J. Biomed. Opt. **30**(2), 023512 (2025)
7. Li, H., et al.: Towards RGB-NIR cross-modality image registration and beyond. arXiv preprint arXiv:2405.19914 (2024)
8. Li, P., MacCormac, O., Shapey, J., Vercauteren, T.: A self-supervised and adversarial approach to hyperspectral demosaicking and RGB reconstruction in surgical imaging. In: 35th British Machine Vision Conference 2024, BMVC 2024, Glasgow, UK, 25–28 November 2024. BMVA (2024). https://papers.bmvc2024.org/0188.pdf

9. Liu, Y., et al.: Motion-boundary-driven unsupervised surgical instrument segmentation in low-quality optical flow (2025). https://arxiv.org/abs/2403.10039
10. Schupper, A.J., et al.: Fluorescence-guided surgery: a review on timing and use in brain tumor surgery. Front. Neurol. **12**, 682151 (2021)
11. Teed, Z., Deng, J.: RAFT: recurrent all-pairs field transforms for optical flow. In: Vedaldi, A., Bischof, H., Brox, T., Frahm, J.-M. (eds.) ECCV 2020, Part II. LNCS, vol. 12347, pp. 402–419. Springer, Cham (2020). https://doi.org/10.1007/978-3-030-58536-5_24
12. Walke, A., Black, D., Valdes, P.A., Stummer, W., König, S., Suero-Molina, E.: Challenges in, and recommendations for, hyperspectral imaging in ex vivo malignant glioma biopsy measurements. Sci. Rep. **13**(1), 3829 (2023)
13. Xie, Y., et al.: Wide-field spectrally resolved quantitative fluorescence imaging system: toward neurosurgical guidance in glioma resection. J. Biomed. Opt. **22**(11), 116006 (2017)
14. Zhai, M., Ni, K., Xie, J., Gao, H.: Cross-modal optical flow estimation via modality compensation and alignment. In: ICASSP 2023-2023 IEEE International Conference on Acoustics, Speech and Signal Processing (ICASSP), pp. 1–5. IEEE (2023)
15. Zhou, S., Tan, W., Yan, B.: Promoting single-modal optical flow network for diverse cross-modal flow estimation. In: Proceedings of the AAAI Conference on Artificial Intelligence, vol. 36, no. 3, pp. 3562–3570 (2022)

Cardio-Respiratory Motion Estimation and Coronary Artery Segmentation for Image-Guided Percutaneous Coronary Intervention

D. China[1], G. Kim[2], N. Iyer[1], R. McGovern[1], A. Uneri[1], and J. Lee[2(✉)]

[1] Biomedical Engineering, Johns Hopkins University, Baltimore, USA
[2] Radiation Oncology and Molecular Radiation Sciences, Johns Hopkins University, Baltimore, USA
junghoon@jhu.edu

Abstract. Image guidance during percutaneous coronary interventions (PCI) can help minimize radiation exposure and contrast use while ensuring procedural safety and efficacy. To support this, this work proposes a framework that leverages a patient-specific cardio-respiratory motion model, optimized intra-procedurally, to enable real-time vessel tracking. The approach is built on: (i) a population-derived motion model capturing cardiac and respiratory dynamics, and (ii) an automated coronary artery segmentation pipeline for both 3D computed tomography angiography (CTA) and 2D x-ray angiography (XA). The motion model integrates cardiac phase and respiratory surrogates, including cycle phase and inhalation/exhalation ratio. To enable training and validation, paired 3D+t CTA and 2D+t XA sequences are synthetically generated using the proposed motion model. Coronary artery segmentation is performed using a dual-convolution-transformer U-Net. The approach was evaluated by comparing the segmented left ventricle across simulated and ground-truth 4D cardiac Magnetic Resonance Angiography datasets, demonstrating volume consistency within the 95% confidence interval. Segmentation achieved high Dice similarity scores: 0.86 ± 0.02 (CTA), 0.98 ± 0.01 (simulated XA), and 0.78 ± 0.01 (real XA). These results validate the accuracy of the synthetic motion simulation and segmentation pipeline. Future steps involve tracking of vessels by estimating patient-specific cardio-respiratory motion by using the proposed population-derived motion and segmented coronary arteries.

Keywords: Motion estimation · coronary artery segmentation · image-guided surgery · surgical navigation · percutaneous coronary interventions

1 Introduction

Pre-operative computed tomography angiography (CTA) is commonly used to visualize cardiac artery structures [1], while real-time tracking is primarily achieved using X-ray coronary angiography (XA) [2–5] during percutaneous coronary interventions (PCI).

D. China and G. Kim—These authors contributed equally to this work.

XA employs 2D fluoroscopy and contrast agents to visualize the arteries during the procedure. The continuous use of fluoroscopy and contrast agents during PCI can increase the risk of radiation-induced skin injury and cancer [6], contrast-induced nephropathy [7], hypotension, and arrhythmias [8]. These risks highlight the need for limiting fluoroscopy time and minimizing contrast use. The PCI procedure is complicated due to the dynamic nature of the coronary artery, which is influenced by both cardiac and respiratory motion. To reduce fluoroscopy time and contrast use during PCI, a patient-specific cardio-respiratory motion model can be employed to track the coronary artery continuously. Numerous studies have proposed various motion simulations, including respiratory motion tracking during image-guided surgery and radiotherapy [9], and cardiac motion estimation using a classical statistical motion model [10], which is built to estimate motion only on the arterial structures, not in the entire cardiac surface. Xie et al. [11] proposed a patient- and segment-specific cardiac motion model for stereotactic radio-ablation of arrhythmias. This technique is limited to breadth-hold cardiac CT only.

Intraoperative vessel tracking during PCI can be achieved by optimizing the parameters of a cardio-respiratory motion model (e.g., derived from population data) via registration between CTA and XA images. Segmenting the arterial structures from both imaging modalities can simplify this registration process. Several studies have aimed to automate this segmentation process. Zeng et al. [12] introduced a two-stage, multi-scale patch fusion approach using 3D U-Net and U-Net++ for coronary artery segmentation in CTA. Cervantes-Sanchez et al. [13] employed a multilayer perceptron with multiscale filter responses (Gaussian matched and Gabor filters) to enhance vessel-like features in XA images. Samuel and Veeramalai [14] proposed VSSC Net, integrating vessel-specific layers into a VGG-16 backbone. However, these methods are typically constrained to either 2D or 3D segmentation. In this study, we utilize a Dual Convolution-Transformer U-Net (DCT-UNet) [15] to segment coronary arteries in both CTA and XA images using modality-specific training. The main contributions of this work are the following:

(1) We propose a cardio-respiratory motion model derived from a set of 4D data using principal component analysis (PCA) and respiratory surrogates. (2) We utilize the cardio-respiratory motion model to generate paired 4D (3D+t) CTA and 2D+t XA images from a patient's 3D CTA. This is particularly important to simulate realistic images with varying motion patterns (useful for any image processing algorithm/model training) as well as to estimate patient-specific motion for PCI. (3) We developed automatic coronary artery segmentation methods for both 3D CTA and 2D XA, thereby enabling the subsequent optimization of motion model parameters for patient-specific motion estimation.

2 Methods

The detailed workflow of the arterial navigation during PCI and simulated paired dataset generation is shown in Fig. 1. This workflow comprises four core components: (1) population-based cardio-respiratory motion estimation, (2) generation of 3D+t CTA and 2D+t XA images from the patient's 3D CTA scan, (3) segmentation of 3D and 2D coronary artery centerline, and (4) 3D-2D registration to optimize the parameters (green arrow in Fig. 1) of the population-based cardio-respiratory motion model to obtain a patient-specific motion model and navigate the coronary arteries. This study focuses on detailing the first three components of architecture. The fourth component, involving motion parameter optimization through 3D–2D registration, is identified as future work.

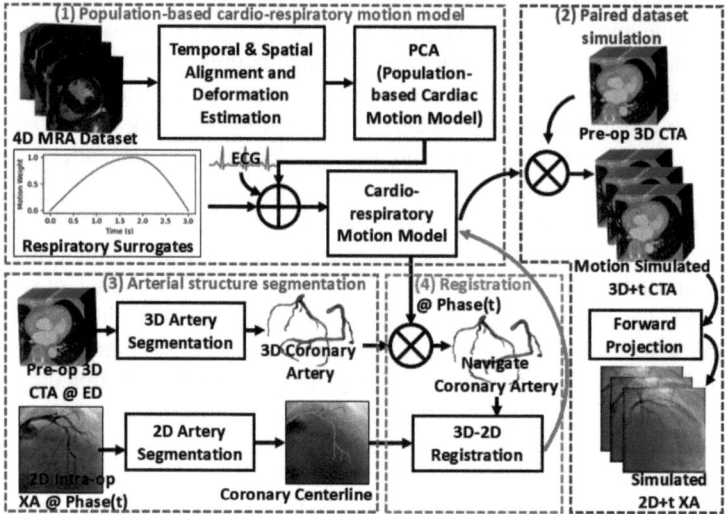

Fig. 1. Overview of the proposed framework for coronary artery navigation during PCI and its utilization for generating paired 3D+t CTA and 2D+t XA datasets.

2.1 Population-Based Cardiac Motion Model

We derive a cardiac motion from the cardiac surface using a publicly available 4D cardiac MRI dataset, "Automated Cardiac Diagnosis Challenge" [16]. This dataset includes 150 patient images acquired at breath-hold with retrospectively or prospectively-gated 4D MRIs. End-diastole (ED) and end-systole (ES) frames were identified in each 4D acquisition, and the masks for the left ventricle (LV), myocardium (MC), and right ventricle (RV) were provided for both ED and ES phases. Each patient's ($p \in [1,150]$) cardiac surface motion (displacement fields), $\mathcal{M}^p$ across all the cardiac phases were estimated using deformable image registration ($\mathfrak{R}$) between each phase's 3D volume ($V_t, t \in [1, T_{RR}]$) and ED phase volume V_{ed}, where T_{RR} is the cardiac cycle R-R interval.

$$\mathcal{M}^p = [\mathfrak{R}(V_{ed}, V_1), \mathfrak{R}(V_{ed}, V_2), \ldots, \mathfrak{R}(V_{ed}, V_{T_{RR}})] \quad (1)$$

In this study, deformable image registration was performed using a voxel-wise displacement field transform with a mean square metric to estimate the dense displacement fields between two phases. The registration refines an initial displacement field transform at multiple levels of resolution with shrink factors and smoothing applied per level. The spatial resolution and origin of each patient's image in the dataset are different. We therefore align all the patients' ED phases V_{ed}^p to the reference ED phase V_{ed}^r using a deformable transform ($\mathfrak{T}$) between masks of LV (l_{lv}), MC (l_{mc}), RV (l_{rv}). Finally, images are cropped around the cardiac surface of the V^r_{ed} for all patient's displacement field ($\mathcal{M}_S$) to get the same volume size across all the $\mathcal{M}_S$.

$$\mathcal{M}^p_S = \mathfrak{T}(\mathcal{M}^p; l^r_{lv}, l^r_{mc}, l^r_{rv}, l^p_{lv}, l^p_{mc}, l^p_{rv}) \tag{2}$$

Approximately 70% of the acquired dataset has 30 temporal phases between the R-R interval (T_{RR}). So, we interpolated the temporal phase of all the $\mathcal{M}^p_S$ to [1, 30]. The interpolation of the temporal alignment between ED to ES and ES to ED was done separately. Finally, the cardiac surface motion ($\mathcal{M}_c$) was estimated using PCA from those Spatio-temporally aligned computed deformation fields and modes of variations φ as

$$\mathcal{M}_c = w \cdot \varphi \tag{3}$$

where w is the weight. In this study, we used 5 principal components that characterized 95% of the population motion. Reference ED phase volume V^r_{ed} is used as mean shape to apply $\mathcal{M}_c$ to characterize the motion.

2.2 Cardio-Respiratory Motion Model Estimation

In this study, we developed a cardio-respiratory motion model by merging the respiratory surrogates along with a population-based cardiac motion model. If f (Hz) is the breathing rate of a patient, $T_{\text{cycle}} = 1/f$ is the duration of one respiratory cycle. For each time point $t \in [1, T_{\text{cycle}}]$, the respiratory motion can be estimated as:

$$\mathcal{M}_r = \begin{cases} A \sin\left(\frac{\pi t/\alpha T_{\text{cycle}}}{2}\right), & \text{if } 1 \leq t \leq \alpha T_{\text{cycle}} \\ A \cos\left(\frac{\pi (t-\alpha T_{\text{cycle}})/(1-\alpha)T_{\text{cycle}}}{2}\right), & \text{if } \alpha T_{\text{cycle}} < t \leq T_{\text{cycle}} \end{cases} \tag{4}$$

where A represents the amplitude of the respiratory motion and α represents the inhalation duration ratio. To capture realistic respiration-induced motion, this motion was applied to various organs using different weights, such as in the soft tissue region (heart chambers, lung, and liver segments) higher weights were applied, whereas in the bony region (ribs, vertebrae, etc.) lower weights were applied. Gaussian filtering was used to smooth the transition of the motion between soft tissue organs. The maximum amplitudes of displacement were set in the antero-posterior and superior-inferior directions. Cardio-respiratory motion model was created by merging the respiratory motion ($\mathcal{M}_r$) with cardiac motion model ($\mathcal{M}_c$). The $\mathcal{M}_c$ assumes a T_{RR} second cardiac cycle from ED-to-ED and T_{cycle} second respiratory cycle, for simulation of cardio-respiratory dynamics. So, the cardio-respiratory motion can be modeled as:

$$\mathcal{M}(t) = \mathcal{M}_c(t') + \mathcal{M}_r(t), \quad t' = t - \left(T_{RR} \left\lfloor \frac{t}{T_{RR}} \right\rfloor\right) \tag{5}$$

2.3 Paired 3D+t CTA and 2D+t XA Generation from 3D CTA

This comprehensive framework of population-based motion ($\mathcal{M}$) estimation enables the simulation of motion-induced 3D+t CTA and 2D+t XA sequences. To integrate cardio-respiratory motion ($\mathcal{M}$) to the patient's 3D CTA, we perform another alignment similar to Eq. (2) between LV, MC, and RV masks of V_{ed}^r and patient's 3D CTA. LV, MC, and RV masks of CTA were generated using "TotalSegmentator" [17]. Then $\mathcal{M}$ was applied to 3D CTA to reconstruct a 3D+t CTA. Digitally reconstructed radiographs were generated from each phase of the 3D+t CTA, resulting in realistic 2D+t XA images. Trilinear interpolation was used to forward project the CTA of each phase, where we set the parameters, including the gantry angle of 0°, source to detector distance of 1200 mm, radiograph image size of 512 × 512, and pixel resolution of 0.5 × 0.5 mm². The motion in the generated 3D+t CTA and 2D+t XA datasets closely mimics real patients' cardio-respiratory motion. These generated datasets are essential as real patients' paired 3D+t CTA and 2D+t XA datasets are not readily available for the development of real-time vessel tracking algorithms (3D-2D image registration) and their validation.

2.4 Coronary Artery Segmentation

We employed the DCT-UNet [15] for coronary artery segmentation. The DCT-UNet was originally developed for organs and tumor segmentation on 3D female pelvic MR images, where it demonstrated excellent performance in capturing complex anatomical structures. We adopted this architecture to address the specific challenges of coronary artery segmentation, leveraging its ability to maintain global vessel connectivity while preserving fine structural details. We extended the application of the DCT-UNet to 2D XA, broadening its utility beyond the original 3D scope.

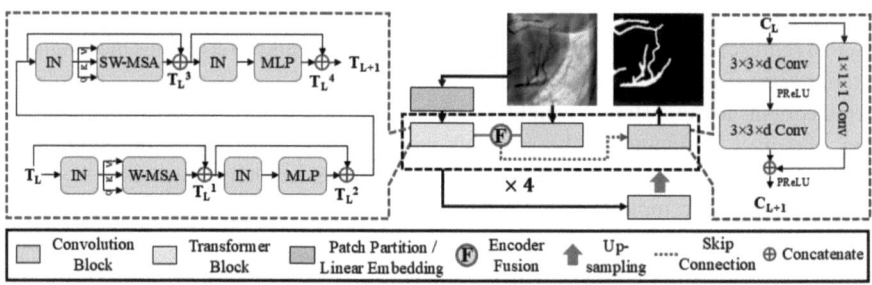

Fig. 2. DCT-UNet for coronary artery segmentation.

The DCT-UNet consists of two main components: a dual-path encoder and a decoder with skip connections (Fig. 2). The dual-path structure combines convolutional and transformer paths to leverage the advantages of both architectures. The convolutional path processes the input image through four sequential convolution blocks, which employ a two-level residual function comprising a 3 × 3 × d convolution, instance normalization (IN), dropout, and parametric rectified linear unit activation (PReLU). The transformer path begins with patch partitioning and linear embedding that convert the spatial information in the input image into a sequence of tokens. These tokenized patches are then fed into a transformer block, which is composed of two consecutive units, each integrating IN, multi-head self-attention (MSA), and a multi-layer perceptron (MLP). To address the intensive computational demands of high-resolution images, the first unit employs window-based MSA (W-MSA), while the second utilizes shifted window-based MSA (SW-MSA) [18]. The feature maps from the two blocks are residually merged in the fusion block, then skip-connected to the decoder.

2.5 Segmentation Model Training

For coronary artery segmentation, we utilized 1,000 3D CTA scans from the ImageCAS dataset [12] and 134 2D XA images from the DCA1 dataset [13]. For 2D arterial structure segmentation, we also generated 100 simulated 2D XA data using 3D CTA data following the method described in Sect. 2.3. In the raw projection dataset (D_r), we applied contrast-limited adaptive histogram equalization (CLAHE) as a post-processing to enhance vessel visibility and improve anatomical structure differentiation, resulting in a contrast-enhanced dataset (D_e). The corresponding segmentation labels were created by projecting the vessel annotations from the 3D CTA data onto the 2D XA plane. The DCT-UNet was trained exclusively on D_e, and evaluated on both D_r and D_e to assess its generalizability across different image contrast conditions. The model was trained on a combined loss function, consisting of Dice and focal losses with weight of 0.7 and 0.3, respectively. The DCT-UNet model was implemented using PyTorch with the MONAI [19] framework and trained on a NVIDIA GeForce RTX A5000. For optimization, we used the AdamW optimizer with an initial learning rate of 0.001 and a weight decay of 0.01, along with a cosine annealing scheduler to gradually reduce the learning rate over epochs.

3 Results

3.1 Cardio-Respiratory Motion Simulation

Figure 3 demonstrates the simulated motion of the right (a-d) and left (e-h) coronary arteries across different phases of the cardio-respiratory cycle in XA. Ventricular volumes (volumetric metrics) and structures (Dice scores) throughout the cardiac cycle were quantitatively compared between the actual patient's 50 testing 4D data sets and the simulated 4D CTA from the cardiac motion model ($\mathcal{M}_c$). Dice scores were computed for the LV for each phase of the cardiac motion, where ED was chosen as the reference phase. Figures 3(i) and 3(j) compare ventricular changes between the left ventricle from the real-patient data (blue curves) and the 3D CTA with the cardiac motion model applied (orange curves) over different phases. The light blue range signifies the 95% confidence interval (CI) around the plotted blue line. The simulated cardiac motion closely resembles and falls within the 95% CI of the real patient cardiac motion, with similar structural and volumetric changes observed from ED-to-ED. This model will be utilized as a prior motion model to estimate patient-specific cardio-respiratory motion by optimizing the prior motion model parameters, such as principal component coefficient (w in Eq. 3), breath rate, respiration cycle, and amplitude displacement. This optimization will be performed using the segmented artery of preoperative CTA and each phase XA segmented coronary artery structure alignment.

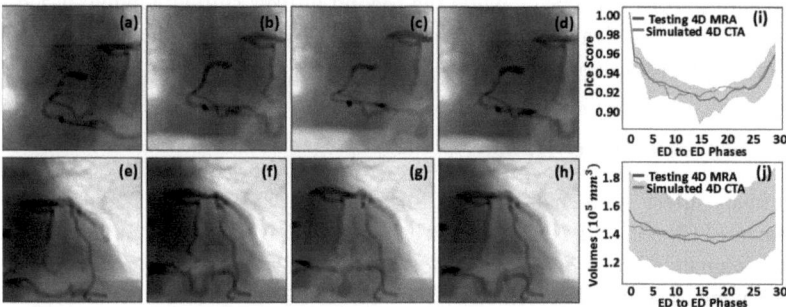

Fig. 3. Simulated XA (2D+t) showing the right (a–d) and left (e–h) coronary arteries. Dice similarity scores (i) and left ventricle volumes (j) over various phases between ED-to-ED were measured in the public dataset (blue) and the simulated 4D CTA (orange). (Color figure online)

3.2 Coronary Artery Segmentation

The performance of coronary artery segmentation was quantitatively evaluated using the dice similarity coefficient (DSC) and the 95th percentile of Hausdorff distance (HD95). DSC quantifies spatial overlap, while HD95 measures boundary discrepancy between prediction and ground truth. In our prior study [20], we applied the DCT-UNet for coronary artery segmentation on 3D CTA, and achieved DSC of 0.86 ± 0.02, HD95 of 0.98 ± 0.98 mm, average HD of 0.20 ± 0.08 mm, and maximum HD of 16.02 ± 11.63 mm, outperforming three existing state-of-the-art models (i.e. baseline model and direct segmentation model in [12], and nnUNet v2 [21]). In this study, we further extended the application of coronary artery segmentation using DCT-UNet to 2D XA. For the simulated dataset, the DCT-UNet was trained, validated, and tested on 80, 10, and 10 D_e images, respectively, and achieved DSC of 0.98 ± 0.01 and HD95 of 0.37 ± 2.02 mm, demonstrating quantitatively excellent performance. Moreover, the trained model was applied to the test D_r images to evaluate its robustness across varying image contrast conditions. We achieved a slightly decreased DSC of 0.97 ± 0.01 due to the lower vessel contrast compared to D_e as shown in Fig. 4. Nonetheless, the performance remains comparable, demonstrating the robustness of the model under non-optimized conditions. Although the HD95 was 10.32, this metric was largely influenced by a few cases (e.g., case in row (b) and column S_r of Fig. 4) with isolated false positives that were detected far from the main vessel region.

While simulated XA images may show different characteristics in noise, artifacts, and contrast dynamics from those in real XA, our approach can generate an arbitrarily large number of realistic XA images with varying characteristics. Such widely variable simulated XA images can help train a model to adapt to a wide range of image quality variability observed in real clinical cases. Furthermore, real 2D+t XA images will be for 3D-2D registration [Fig. 1(4)] instead of the simulated images in practical clinical. Therefore, we additionally evaluated the DCT-UNet on the real patients' DCA1 dataset along with the simulated dataset. We (indirectly) compare the segmentation performance with state-of-the-art methods as summarized in Table 1. The DCT-UNet can segment the ground truth region accurately, achieving the highest sensitivity, while other metric values are comparably similar to the other methods. Considering that the ground truth coronary artery segmentation in DCA1 data is not accurate in some test cases [e.g., Fig. 4 (c-d) GT column], DCT-UNet performance is considered among the state-of-the-art. The computation time for arterial structure segmentation per XA image was 1.20 s.

Table 1. Comparison of XA image segmentation results based on Dice (DSC), sensitivity (Sen), and specificity (Spe).

Method	DSC	Sen	Spe
Cervantes-Sanchez et al. [13]	0.69	0.64	0.99
Samuel et al. [14]	–	0.77	0.98
Li et al. [22]	0.73	0.83	–
Xu et al. [23]	0.77	0.84	0.99
Kus and Kiraz [24]	0.78	0.81	0.99
DCT-UNet	0.78	0.88	0.98

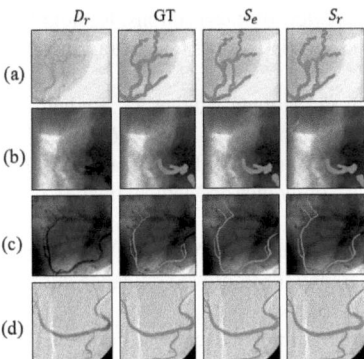

Fig. 4. Segmentation of the original (S_r) and enhanced (S_e) XA images. (a-b) for the simulated XA image, (c-d) are for real patient XA image.

4 Conclusions

In this study, we proposed a cardio-respiratory motion model that can mimic real patients' cardio-respiratory motion. Paired datasets were simulated using the proposed motion model, which will be essential for the validation of 3D-2D image registration, as real patients' datasets are not readily available. Furthermore, a DCT-UNet is employed to achieve accurate segmentation of arterial structures from both 3D and 2D images. Ongoing work focuses on integrating these components into a comprehensive pipeline for 3D-2D registration, enabling estimation of patient-specific motion in both simulated and clinical settings.

Acknowledgements. This study was supported by LN Robotics, Inc. and the Ministry of Trade, Industry and Energy (MOTIE), Korea, under the "Global Industrial Technology Cooperation Center (GITCC) program" supervised by the Korea Institute for Advancement of Technology (KIAT, Task No. P0028454).

References

1. Hennessey, B., Vera-Urquiza, R., Mejia-Renteria, H., Gonzalo, N., Escaned, J.: Contemporary use of coronary computed tomography angiography in the planning of percutaneous coronary intervention. Int. J. Cardiovasc. Imaging **36**, 2441–2459 (2020)
2. Matl, S., Brosig, R., Baust, M., Navab, N., Demirci, S.: Vascular image registration techniques: a living review. Med. Image Anal. **35**, 1–17 (2017)
3. Yoon, S., Yoon, C.H., Lee, D.: Topological recovery for non-rigid 2D/3D registration of coronary artery models. Comput. Methods Programs Biomed. **200**, 105922 (2021)
4. Kim, H.R., Kang, M.S., Kim, M.H.: Non-rigid registration of vascular structures for aligning 2D X-ray angiography with 3D CT angiography. In: Advances in Visual Computing, vol. 8887 (2014)
5. Chen, Y., et al.: GVM-Net: a GNN-based vessel matching network for 2D/3D non-rigid coronary artery registration. IEEE Trans. Med. Imaging (2025)
6. Burke, G., Faithfull, S., Probst, H.: Radiation induced skin reactions during and following radiotherapy: a systematic review of interventions. Radiography **28**(1), 232–239 (2022)
7. Toprak, O.: Conflicting and new risk factors for contrast induced nephropathy. J. Urol. **178**(6), 2277–2283 (2007)
8. Roobottom, C.A., Mitchell, G., Morgan-Hughes, G.: Radiation-reduction strategies in cardiac computed tomographic angiography. Clin. Radiol. **65**(11), 859–867 (2010)
9. Klinder, T., Lorenz, C., Ostermann, J.: Prediction framework for statistical respiratory motion modeling. In: International Conference on Medical Image Computing and Computer-Assisted Intervention, pp. 327–334 (2010)
10. Baka, N., Lelieveldt, B.P.F., Schultz, C., Niessen, W., van Walsum, T.: Respiratory motion estimation in x-ray angiography for improved guidance during coronary interventions. Phys. Med. Biol. **60**(9), 3617 (2015)
11. Xie, J., et al.: Electrocardiogram-gated cardiac computed tomography-based patient-and segment-specific cardiac motion estimation method in stereotactic arrhythmia radioablation for ventricular tachycardia. Phys. Imaging Radiat. Oncol., 100700 (2025)
12. Zeng, A., et al.: ImageCAS: a large-scale dataset and benchmark for coronary artery segmentation based on computed tomography angiography images. Comput. Med. Imaging Graph. **109**, 102287 (2023)
13. Cervantes-Sanchez, F., Cruz-Aceves, I., Hernandez-Aguirre, A., Hernandez-Gonzalez, M.A., Solorio-Meza, S.E.: Automatic segmentation of coronary arteries in X-ray angiograms using multiscale analysis and artificial neural networks. Appl. Sci. **9**(24), 5507 (2019)
14. Samuel, P.M., Veeramalai, T.: VSSC Net: vessel specific skip chain convolutional network for blood vessel segmentation. Comput. Methods Programs Biomed. **198**, 5769 (2021)

15. Kim, G., et al.: Dual convolution-transformer UNet (DCT-UNet) for organs at risk and clinical target volume segmentation in MRI for cervical cancer brachytherapy. Phys. Med. Biol. **69**(21), 215014 (2024)
16. Bernard, O., et al.: Deep learning techniques for automatic MRI cardiac multi-structures segmentation and diagnosis: is the problem solved? IEEE Trans. Med. Imaging **37**(11), 2514–2525 (2018)
17. Wasserthal, J., et al.: TotalSegmentator: robust segmentation of 104 anatomic structures in CT images. Radiol. Artif. Intell. **5**(5), 230024 (2023)
18. Liu, Z., et al.: Swin transformer: Hierarchical vision transformer using shifted windows. In: Proceedings of the IEEE/CVF International Conference on Computer Vision, pp. 10012–10022 (2021)
19. Cardoso, M.J., et al.: MONAI: an open-source framework for deep learning in healthcare. arXiv preprint arXiv:2211.02701, (2022)
20. McGovern, R., Kim, G., Lee, J.: Coronary artery segmentation with dual convolution-transformer U-Net for cardiovascular interventions. In: SPIE Medical Imaging 2025: Clinical and Biomedical Imaging, vol. 13410, pp. 123–127 (2025)
21. Isensee, F., Jaeger, P.F., Kohl, S.A., Petersen, J., Maier-Hein, K.H.: NnU-Net: a self-configuring method for deep learning-based biomedical image segmentation. Nat. Methods **18**(2), 203–211 (2021)
22. Li, Z., Zhang, H., Li, Z., Ren, Z.: Residual-attention UNet++: a nested residual-attention U-net for medical image segmentation. Appl. Sci. **12**(14), 7149 (2022)
23. Xu, H., Wu, Y.: G2ViT: graph neural network-guided vision transformer enhanced network for retinal vessel and coronary angiograph segmentation. Neural Netw. **176**, 106356 (2024)
24. Kuş, Z., Kiraz, B.: Evolutionary architecture optimization for retinal vessel segmentation. IEEE J. Biomed. Health Inform. **27**(12), 5895–5903 (2023)

Arachnoid Membrane Segmentation in Intraoperative Microscopic MVD Surgery Scenes

Jinhee Lee[2,3], Hwanhee Lee[1,2,3], Jay J. Park[3,4], Jeong Woo Ahn[2], Jong Yun Kwon[2], Ciara McMahon[3,4], Julia Lewandowski[3], Seohee Park[3], Sanghoon Lee[1], and Vivek P. Buch[3(✉)]

[1] Department of Neurosurgery, Bongseng Memorial Hospital, Busan, South Korea
[2] Bongseng AI Lab (BAIL), Bongseng Memorial Hospital, Busan, South Korea
[3] Surgical Innovation and Machine Interfacing (SIMI) Laboratory, Department of Neurosurgery, Stanford University, Stanford, CA, USA
vpbuch@stanford.edu
[4] Edinburgh Medical School, The University of Edinburgh, Edinburgh, UK

Abstract. Microvascular decompression (MVD) is a neurosurgical procedure to treat cranial nerve compression syndromes such as trigeminal neuralgia and hemifacial spasm. The arachnoid membrane (AM) is a thin, transparent meningeal layer that adheres to or covers neurovascular structures and must be carefully dissected to access the surgical site during MVD surgery. Proper AM dissection is essential for visualizing the operative field and ensuring safe vessel and nerve manipulation. However, AM dissection is technically challenging due to its poor contrast with surrounding tissues and close adherence to critical neurovascular structures. To address this, we propose the first dedicated study on AM segmentation from operative MVD videos. We introduce a high-quality, expert-annotated dataset focusing on AM in the cerebellopontine angle and train a segmentation model with a task-specific loss function to improve AM segmentation. Our results demonstrate that the proposed loss function improves AM segmentation performance by 7.35% in IoU over the baseline, enabling reliable segmentation despite the membraneâĂŹs transparency and intraoperative variability. This work lays the foundation for automated AM recognition in surgical environments and provides a valuable resource for AM dissection and surgical decision-making.

Keywords: Microvascular decompression (MVD) · Arachnoid Membrane · Deep Learning · Semantic Segmentation · Surgical AI

1 Introduction

Microvascular decompression (MVD) is performed to relieve neurovascular compression syndromes, most notably trigeminal neuralgia (TN) and hemifacial

---
J. Lee and H. Lee— Equal Contribution.

© The Author(s), under exclusive license to Springer Nature Switzerland AG 2026
Q. Dou et al. (Eds.): COLAS 2025, LNCS 16298, pp. 168–177, 2026.
https://doi.org/10.1007/978-3-032-09784-2_17

spasm (HFS), by repositioning or separating the offending vessels, typically the superior cerebellar artery (SCA), anterior inferior cerebellar artery (AICA), or posterior inferior cerebellar artery (PICA), from the trigeminal nerve (cranial nerve V) in TN or the facial nerve (cranial nerve VII) in HFS [9,16]. Sustained neurovascular contact can cause focal demyelination and abnormal signal conduction, typically at transitional zones such as the root entry zone in TN or the root exit zone in HFS, leading to neuropathic pain or involuntary movements [7]. These procedures are performed in anatomically dense operative fields, where critical neurovascular structures must be clearly visualized before safe decompression can proceed [10].

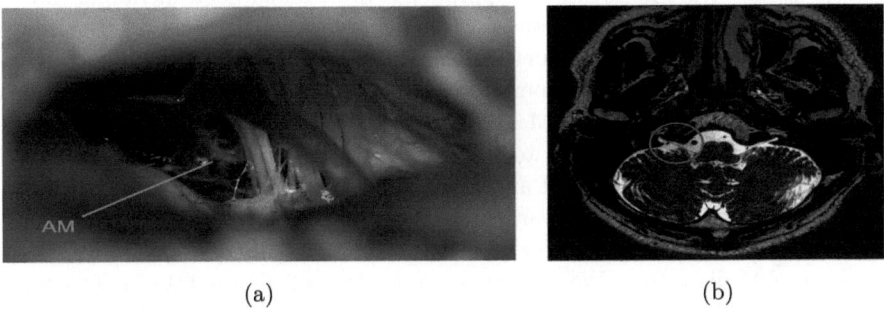

Fig. 1. Intraoperative microscopic image of the corresponding area, where the AM is clearly visible at the location indicated by the green line (a). Preoperative CISS MRI scan showing no visible AM in the region marked by a red circle (b). (Color figure online)

One of the main challenges in achieving clear surgical exposure is the arachnoid membrane (AM), a thin, avascular, and translucent meningeal layer that often adheres to nerves, vessels, and surrounding tissues [1,3]. To ensure both the safety and completeness of the procedure, the AM must be carefully dissected along appropriate anatomical planes [17]. This requires accurately identifying the extent and boundaries of the membrane, particularly where it partially covers or adheres to critical neurovascular structures [6]. However, AM is not discernible on the corresponding preoperative magnetic resonance (MR) images. This limitation is due to the AM's sub-voxel thickness (approximately 49 μm) and its signal similarity to cerebrospinal fluid, which together cause partial volume effects that make it effectively invisible on MRI [3,12,15,20]. This invisibility on preoperative imaging necessitates that AM dissection be performed entirely based on intraoperative visual assessment. The AM is clearly visualized under the surgical microscope in Fig. 1 (a), and it is not visible in the corresponding preoperative MRI of the same region, as shown in Fig. 1(b).

As illustrated in Fig. 2, the AM (green) initially obscures the CN5 (red) and the offending artery (yellow), which become visible only after membrane dissection [10,17]. Surgeons must delineate and dissect the AM in real time under the surgical microscope or endoscope, relying solely on visual cues such as its reflectivity, subtle motion, and contrast with adjacent structures [1,15]. Inaccurate dissection planes may obscure critical anatomy or cause neurovascular injury [17], often necessitating cautious surgical progression that prolongs operative time and increases procedural risk [18]. These challenges highlight the need for intelligent systems that can assist with intraoperative AM localization and dissection guidance, either by enhancing the surgeonâĂŹs visual perception or through future autonomous technologies [8,11].

The application of AI to surgical vision has rapidly expanded, with growing emphasis on real-time perception, semantic segmentation of intraoperative scenes, and building the concept of the Surgeon-Machine Interface (SMI) [19]. Several studies have applied such approaches to brain surgery, including MVD, involving real-time segmentation of cranial nerves and arteries [2], boundary refinement of neurovascular structures using EnsembleEdgeFusion [5], and absolute depth estimation in intraoperative MVD scenes [13]. Despite its ubiquitous presence in brain surgeries, AM has remained relatively unaddressed in surgical AI research, probably due to its ambiguous visual boundaries and signal similarity to cerebrospinal fluid and adjacent tissues [21].

In this paper, we introduce arachnoid membrane segmentation in real-world MVD surgical scenes, laying the groundwork for future AI-assisted surgical tasks involving AM dissection. We construct a new dataset of annotated microscopic images from MVD procedures, with pixel-level AM labels manually created by expert neurosurgeons. To improve performance on this visually subtle target, we propose a query-wise AM loss that enhances segmentation by focusing supervision on AM-relevant regions. Moreover, we explore the clinical relevance of AM segmentation and its potential integration into AI-assisted surgical workflows.

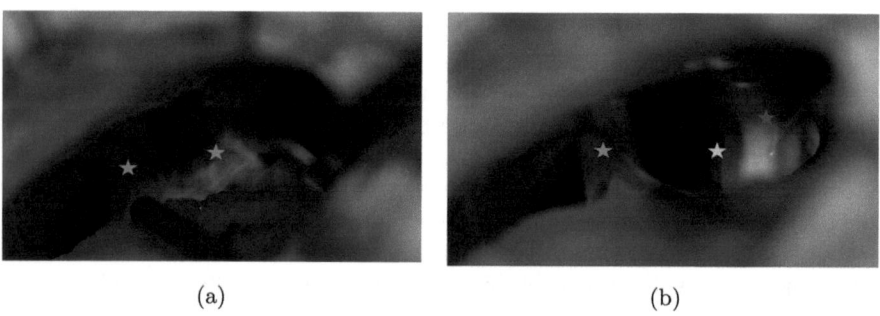

Fig. 2. Intraoperative microscopic images before (a) and after (b) AM dissection during MVD. In (a), the AM (green) overlies the surgical field, and only the CN8 (orange) is visible. After dissection in (b), the AM has been dissected, revealing the CN5 (red) and the offending artery (yellow). (Color figure online)

2 Methodology

In this section, we present the overall methodology of our approach. We first introduce the newly constructed MVD dataset focused on AM regions. Next, we describe the baseline segmentation model architecture based on Mask2Former [4]. Lastly, we explain our proposed Query-wise AM Supervision Loss, which is designed to enhance class-specific segmentation performance.

2.1 Dataset Configuration

We constructed a microscopic MVD segmentation dataset of arachnoid membrane and adjacent anatomical structures. This dataset includes 7 patients who underwent MVD surgeries at Bongseng Memorial Hospital (BMH) performed by a single neurosurgeon between 2020 and April 2025. 5 patients were diagnosed with TN, and 2 patients with HFS. Patient ages ranged from 52 to 84 years, with 5 females and 2 males. The pathology was left-sided in 4 cases and right-sided in 3. The study was approved by the Institutional Review Board (IRB) under protocol number BSIRB-2024-008.

Surgeries were recorded using high-resolution surgical microscopes TIVATO 700 and Vario S88 (Zeiss, Jena, Germany) with videos captured at 1080p resolution and 30 FPS. For each case, representative video segments were selected based on the presence of AM dissection and neurovascular manipulation. A total of 554 frames were manually annotated by neurosurgeons, and all surgical videos were de-identified to ensure patient anonymity.

Annotations were performed based on 40 anatomical classes to capture detailed neurovascular structures, which were subsequently consolidated into 10 classes for AM segmentation by merging related structures to enhance surgical relevance and mitigate class imbalance. Cranial nerves (including CN5, CN7, CN8, CN9, and CN10) were grouped into a single CN class. Vascular structures, consisting of arteries such as AICA, PICA, and SCA, veins including the SPV, and small-caliber vessels, were consolidated into a unified vessel class. Major anatomical structures, including the cerebellum, brainstem, and tentorium, were grouped into a generalized tissue class. Tools used in surgery were categorized into surgical instruments and surgical materials. Surgical instruments included devices such as suction tips and bipolar forceps, while surgical materials included items such as teflon and cottonoids.

To enhance the modelâĂŹs capacity to capture the nuanced appearance and interaction of the AM, we introduced AM-associated interaction labels. AM indicates regions where the membrane is not attached to other structures and represents the pure form of the arachnoid membrane. Depending on the attached structure, CN-AM, vessel-AM, and tissue-AM labels are assigned accordingly.

2.2 Baseline Architecture

We adopt Mask2Former [4] as our baseline segmentation architecture due to its strong performance across various semantic segmentation tasks. Mask2Former consists of a hierarchical backbone that extracts multi-scale image features, a pixel decoder that hierarchically upsamples high-resolution pixel embeddings from low-level features, and a transformer decoder that jointly predicts class labels and corresponding segmentation masks using learnable query embeddings and pixel embeddings. As the hierarchical backbone, we utilize the Swin Transformer [14], which provides robust hierarchical visual representations suitable for dense prediction. The detailed architecture of Mask2Former is depicted in Fig. 3.

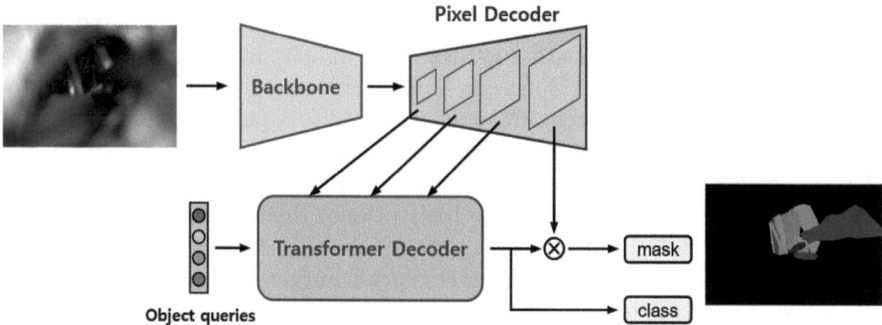

Fig. 3. Illustration of the baseline Mask2Former architecture, composed of a backbone, a pixel decoder, and a transformer decoder that predicts masks via masked attention with learnable object queries.

2.3 Query-Wise AM Supervision Loss

We introduce a query-wise AM supervision loss (AM loss) to enhance segmentation sensitivity for AM regions. Due to the translucent and structurally ambiguous nature of AM, conventional loss terms often fail to provide sufficient supervision for this class. To address this, our AM loss applies direct supervision between the predicted masks and the ground-truth AM mask, encouraging at least one query embedding to specialize in AM-specific features. To focus the supervision on the most informative candidates, we compute the loss only over the top-K queries whose predicted masks exhibit the highest similarity to the AM region. The AM loss is defined as:

$$\mathcal{L}_{\text{am}} = \frac{1}{K} \sum_{q \in \mathcal{T}} \left(1 - \frac{2 \cdot \sum \hat{y}_q \cdot y_{\text{am}}}{\sum \hat{y}_q + \sum y_{\text{am}} + \epsilon} \right) \quad (1)$$

where $\mathcal{T}$ denotes the indices of the top-K queries ranked by overlap with the AM mask, $\hat{y}_q$ denotes the predicted mask from the q-th query in the final decoder layer, y_{am} is the binary ground-truth mask for the AM region, and ϵ is a small constant added for numerical stability.

The overall training objective integrates the standard Mask2Former loss, which consists of a classification loss $\mathcal{L}_{\text{cls}}$, a binary cross-entropy loss $\mathcal{L}_{\text{ce}}$, and a Dice loss $\mathcal{L}_{\text{dice}}$, with the proposed AM loss $\mathcal{L}_{\text{am}}$:

$$\mathcal{L}_{\text{m2f}} = \lambda_{\text{cls}} \cdot \mathcal{L}_{\text{cls}} + \lambda_{\text{ce}} \cdot \mathcal{L}_{\text{ce}} + \lambda_{\text{dice}} \cdot \mathcal{L}_{\text{dice}} \tag{2}$$

$$\mathcal{L}_{\text{total}} = \mathcal{L}_{\text{m2f}} + \lambda_{\text{am}} \cdot \mathcal{L}_{\text{am}} \tag{3}$$

where λ_{cls}, λ_{ce}, λ_{dice}, and λ_{am} are scalar weights for each loss component.

3 Experiments

3.1 Implementation Details

Our method was implemented using Mask2Former in the Detectron2 framework and trained on 2 NVIDIA RTX 4090 GPUs. We used a Swin-Large backbone pretrained on ImageNet-22K with a patch size of 4 and a window size of 12. The model architecture includes a multi-scale deformable attention pixel decoder and a 6-layer masked transformer decoder with 8 attention heads. Optimization was performed using AdamW with a base learning rate of 1e-4, weight decay of 0.05, and a cosine learning rate schedule over 160k iterations. Input images had a resolution of 1920 × 1080, and data augmentation included random resizing, 512 × 512 cropping, color jittering, and horizontal flipping.

3.2 Experimental Results

Quantitative Evaluation. We evaluated the effectiveness of the proposed AM loss on our private AM-MVD dataset using class-wise Intersection over Union (cIoU) and mean IoU (mIoU) as evaluation metrics. Table 1 presents a quantitative comparison between the baseline Mask2Former and the model trained with the proposed AM loss. The AM loss significantly improves segmentation performance on pure AM regions, increasing the IoU for the AM class from 21.09 to 28.44. The overall mIoU also increases slightly, from 33.65 to 34.40, although minor performance drops are observed in certain non-AM classes such as CN and Surgical Instrument.

Table 1. Quantitative comparison of semantic segmentation performance on the AM-MVD dataset showing class-wise IoU and mean IoU for the baseline Mask2Former and our method with AM Loss

Method	Background	CN	Vessel	Surgical Instrument	Surgical Materials	AM	Tissue	CN-AM	Vessel-AM	Tissue-AM	mIoU
Mask2Former (baseline)	89.11	44.91	50.53	26.12	35.51	21.09	29.40	10.37	15.80	13.68	**33.65**
Mask2Former (AM loss)	89.68	39.64	57.27	23.76	39.27	**28.44**	27.52	9.30	23.63	5.48	**34.40**

Qualitative Results. Figure 4 illustrates qualitative comparisons between the original Mask2Former and the same model trained with the proposed AM loss.

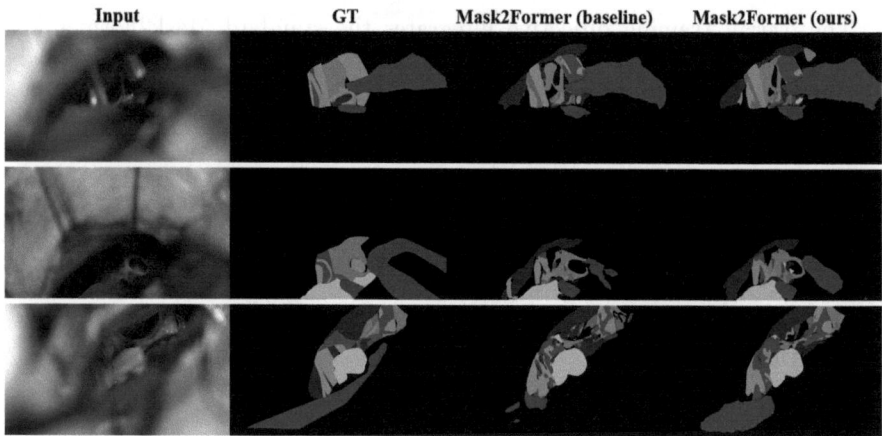

Fig. 4. Comparison of the baseline and AM loss models in MVD scenes. The AM loss model shows improved segmentation of subtle AM regions. Colors: background (black), CN (sky blue), vessels (orange), surgical instruments (dark red), surgical materials (light green), AM (pink), tissue (blue), CN-AM (blue-green), vessel-AM (gold), and tissue-AM (green). (Color figure online)

The results show that incorporating the AM loss makes the model more sensitive to AM regions, leading to more precise segmentation in areas that are often missed by the baseline model. In addition, our method demonstrates strong segmentation performance in complex surgical scenes from MVD procedures, where multiple anatomical structures, such as CNs, vessels, surgical instruments, and AM regions, are densely packed and frequently overlapping. These visual results indicate that the model is capable of clearly delineating such components, highlighting its robustness in real-world intraoperative conditions.

4 Discussion

Our study introduces novel classes of AI assistance that focus on surgically important yet visually ambiguous structures. Segmentation performance on the AM notably improved when membrane-specific visual features were explicitly supervised using the proposed AM loss. The IoU for the AM class increased by 7.35%, and the model became more sensitive to its boundaries and more reliable in distinguishing it from adjacent anatomical entities. However, the model exhibited slight performance decline in other classes, including CN, surgical instruments, and composite regions such as CN-AM and Tissue-AM. These declines are likely attributable to the model's increased emphasis on pure AM regions, which resulted in relatively reduced supervision for overlapping anatomical configurations. Despite these trade-offs, the enhanced sensitivity to membrane boundaries offers meaningful support for advancing AM dissection research and provides support for intraoperative surgical decision-making.

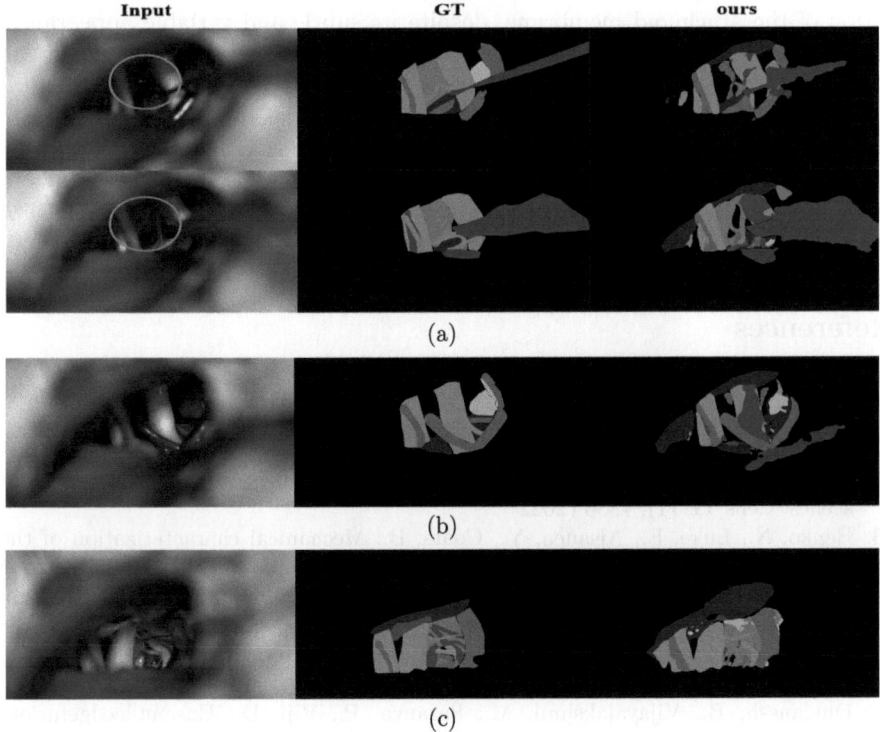

Fig. 5. Representative failure cases in AM segmentation. (a) shows reduced segmentation performance in transparent AM regions. (b) illustrates confusion with fluid-covered or highly reflective cranial nerve surfaces. (c) presents failure to distinguish composite configurations.

Limitations. The transparency of the AM can vary depending on lighting conditions, the presence of surrounding fluid, and membrane thickness. As shown in Fig. 5(a), AM regions with reduced thickness and cerebrospinal fluid fluctuation were segmented less accurately than thicker, more opaque regions. In Fig. 5(b), moisture-covered or brightly illuminated CN surfaces were sometimes misclassified as CN5-AM, reflecting the model's sensitivity to surface reflectance. Figure 5(c) further shows the model's difficulty in identifying composite configurations, such as AM encompassing CN5 (CN-AM), which were often simplified into standalone AM predictions. This tendency highlights persistent limitations in segmenting complex AM configurations.

5 Conclusion

In this work, we introduced the first AM segmentation framework for real-world MVD surgical scenes by constructing a pixel-level annotated dataset and proposing a query-wise AM supervision loss. Our results demonstrate improved segmentation of the arachnoid membrane, despite its subtle and variable appearance. However, challenges remain in generalizing to membrane appearances that are transparent, moist, or affected by lighting conditions. To address these limitations, we plan to improve the robustness of the model to visual variability in the appearance of AM. We are also actively expanding our dataset to include more diverse surgical cases and anatomical variations. In parallel, our goal is to extend this work toward an SMI that not only segments the AM but also supports intraoperative decision making for safe and effective arachnoid dissection.

References

1. Adeeb, N., et al.: The intracranial arachnoid mater: a comprehensive review of its history, anatomy, imaging, and pathology. Childs Nerv. Syst. **29**, 17–33 (2013)
2. Bai, R., Liu, X., Jiang, S., Sun, H.: Deep learning based real-time semantic segmentation of cerebral vessels and cranial nerves in microvascular decompression scenes. Cells **11**(11), 1830 (2022)
3. Benko, N., Luke, E., Alsanea, Y., Coats, B.: Mechanical characterization of the human pia-arachnoid complex. J. Mech. Behav. Biomed. Mater. **120**, 104579 (2021)
4. Cheng, B., Misra, I., Schwing, A.G., Kirillov, A., Girdhar, R.: Masked-attention mask transformer for universal image segmentation. In: Proceedings of the IEEE/CVF Conference on Computer Vision and Pattern Recognition, pp. 1290–1299 (2022)
5. Dhiyanesh, B., Vijayalakshmi, M., Saranya, P., Viji, D.: Ensembleedgefusion: advancing semantic segmentation in microvascular decompression imaging with innovative ensemble techniques. Sci. Rep. **15**(1), 1–22 (2025)
6. Graffeo, C.S., Peris-Celda, M., Perry, A., Carlstrom, L.P., Driscoll, C.L., Link, M.J.: Anatomical step-by-step dissection of complex skull base approaches for trainees: surgical anatomy of the retrosigmoid approach. J. Neurol. Surgery Part B: Skull Base **82**(03), 321–332 (2021)
7. Haller, S., Etienne, L., Kövari, E., Varoquaux, A., Urbach, H., Becker, M.: Imaging of neurovascular compression syndromes: trigeminal neuralgia, hemifacial spasm, vestibular paroxysmia, and glossopharyngeal neuralgia. Am. J. Neuroradiol. **37**(8), 1384–1392 (2016)
8. Iftikhar, M., Saqib, M., Zareen, M., Mumtaz, H.: Artificial intelligence: revolutionizing robotic surgery. Annals Med. Surgery **86**(9), 5401–5409 (2024)
9. Jannetta, P.J.: Observations on the etiology of trigeminal neuralgia, hemifacial spasm, acoustic nerve dysfunction and glossopharyngeal neuralgia. definitive microsurgical treatment and results in 117 patients. Neurochirurgia **20**(05), 145–154 (1977)
10. Kim, B.T.: Significance of arachnoid dissection to obtain optimal exposure of lower cranial nerves and the facial nerve root exit zone during microvascular decompression surgery. J. Korean Neurosurgical Soc. **55**(1), 64 (2014)
11. Kunz, C., et al.: Autonomous planning and intraoperative augmented reality navigation for neurosurgery. IEEE Trans. Med. Robot. Bionics **3**(3), 738–749 (2021)
12. Laitt, R., Mallucci, C., Jaspan, T., McConachie, N., Vloeberghs, M., Punt, J.: Constructive interference in steady-state 3d fourier-transform mri in the management of hydrocephalus and third ventriculostomy. Neuroradiology **41**, 117–123 (1999)

13. Lee, J., et al.: Intraoperative absolute depth estimation in mvd surgery. In: 2025 IEEE 38th International Symposium on Computer-Based Medical Systems (CBMS), pp. 341–342. IEEE Computer Society (2025)
14. Liu, Z., et al.: Swin transformer: hierarchical vision transformer using shifted windows. In: Proceedings of the IEEE/CVF International Conference on Computer Vision, pp. 10012–10022 (2021)
15. Lu, S., Brusic, A., Gaillard, F.: Arachnoid membranes: crawling back into radiologic consciousness. Am. J. Neuroradiol. **43**(2), 167–175 (2022)
16. McLaughlin, M.R., Jannetta, P.J., Clyde, B.L., Subach, B.R., Comey, C.H., Resnick, D.K.: Microvascular decompression of cranial nerves: lessons learned after 4400 operations. J. Neurosurg. **90**(1), 1–8 (1999)
17. Miyake, S., Suenaga, J., Nakamura, T., Akimoto, T., Suzuki, R., Ohtake, M., Takase, H., Tateishi, K., Shimizu, N., Murata, H., et al.: Practical arachnoid anatomy for the technical consideration of galen complex dissection: Cadaveric and clinical evaluation. World neurosurgery **151**, e372–e378 (2021)
18. Okon, I.I., et al.: Microvascular decompression: a contemporary update. BMC Surg. **25**(1), 20 (2025)
19. Park, J.J., et al.: Developing the surgeon-machine interface: using a novel instance-segmentation framework for intraoperative landmark labelling. Front. Surg. **10**, 1259756 (2023)
20. Suzuki, H., Mikami, T., Iwahara, N., Akiyama, Y., Wanibuchi, M., Komatsu, K., Yokoyama, R., Hirano, T., Hosoda, R., Horio, Y., et al.: Aging-associated inflammation and fibrosis in arachnoid membrane. BMC Neurol. **21**, 1–9 (2021)
21. Ungureanu, G., Florian, A., Florian, S.I.: The impact of arachnoid structures on skull-base meningioma surgical management: a radiological analysis and narrative review. J. Med. Life **17**(7), 682 (2024)

DARIL: When Imitation Learning Outperforms Reinforcement Learning in Surgical Action Planning

Maxence Boels[✉], Harry Robertshaw, Thomas C. Booth, Prokar Dasgupta, Alejandro Granados, and Sebastien Ourselin

Surgical and Interventional Engineering, King's College London, London, UK
maxence.boels@kcl.ac.uk

Abstract. Surgical action planning requires predicting future instrument-verb-target triplets for real-time assistance. While teleoperated robotic surgery provides natural expert demonstrations for imitation learning (IL), reinforcement learning (RL) could potentially discover superior strategies through exploration. We present the first comprehensive comparison of IL versus RL for surgical action planning on CholecT50. Our Dual-task Autoregressive Imitation Learning (DARIL) baseline achieves 34.6% action triplet recognition mAP and 33.6% next frame prediction mAP with smooth planning degradation to 29.2% at 10-second horizons. We evaluated three RL variants: world model-based RL, direct video RL, and inverse RL enhancement. Surprisingly, all RL approaches underperformed DARIL—world model RL dropped to 3.1% mAP at 10s while direct video RL achieved only 15.9%. Our analysis reveals that distribution matching on expert-annotated test sets systematically favors IL over potentially valid RL policies that differ from training demonstrations. This challenges assumptions about RL superiority in sequential decision making and provides crucial insights for surgical AI development.

Keywords: Surgical Action Planning · Imitation Learning · Reinforcement Learning · Temporal Planning · Surgical AI

1 Introduction

Surgical action planning, predicting future instrument-verb-target relationships in surgical videos, represents a critical component for real-time surgical assistance systems. Accurate future prediction is essential for enabling proactive surgical guidance, reducing surgeon cognitive load, and facilitating autonomous robotic assistance in complex procedures. While prior work has predominantly focused on recognition tasks [10,11,13], prospective action planning presents unique challenges requiring multi-horizon prediction capabilities under safety-critical constraints.

The fundamental question for surgical AI systems is the optimal learning paradigm: should systems learn through imitation of expert demonstrations (IL) or through trial-and-error optimization via reinforcement learning (RL) [2]? Teleoperated robotic surgery provides natural access to expert demonstrations, making IL attractive. However, RL could theoretically discover strategies beyond expert-level performance through exploration. Recent work in endovascular surgery has demonstrated RL's potential for autonomous navigation tasks, with successful applications in mechanical thrombectomy achieving high success rates while maintaining safety constraints [15,16]. More broadly, RL world models in particular have shown that single configurations with no hyperparameter tuning can outperform specialized methods across diverse benchmark tasks, complete farsighted tasks such as collecting diamonds in Minecraft without human data or curricula, and capture expectations of future events during autonomous driving [3–5].

Recent advances in surgical gesture prediction [18,19] and vision transformers for surgical analysis [6,7] have shown promise, while long-term workflow prediction [1] demonstrates the potential for anticipatory systems. Yet the comparative effectiveness of IL versus RL for surgical action planning remains unexplored.

This work addresses this gap through the first systematic comparison of IL and RL approaches for surgical action planning. Using the CholecT50 dataset [13], we evaluate recognition accuracy and planning capability across multiple time horizons.

Contributions: (1) *First systematic IL vs RL comparison*: Comprehensive evaluation addressing fundamental methodological questions in surgical AI. (2) *Surprising negative results*: RL methods consistently underperform IL with world model RL dropping to 3.1% mAP vs 29.2% for our Dual-task Autoregressive Imitation Learning (DARIL) at 10s planning. (3) *Novel DARIL architecture*: Dual-task autoregressive approach maintaining robust temporal consistency (34.6% action triplet recognition degrading smoothly to 29.2% at 10s). (4) *Evaluation bias insights*: Analysis of how expert demonstration alignment systematically favors IL over valid RL policies.

2 Methods

2.1 Problem Formulation

Given surgical video frames $\{f_1, f_2, ..., f_t\}$, we predict future action triplets $\{a_{t+1}, a_{t+2}, ..., a_{t+H}\}$ where H represents the planning horizon. Each triplet $a_i = (I_i, V_i, T_i)$ consists of instrument, verb, and target components from predefined vocabularies. Importantly, each frame can contain multiple simultaneous actions (0–3 per frame) due to multiple robotic arms with different instruments, making the problem highly sparse with 100 distinct action classes total. We evaluate both current action recognition and future action prediction with single-step ($H = 1$) next frame prediction and multi-step prediction ($H > 1$) for planning assessment across longer horizons.

2.2 Dataset

We use CholecT50 [13], containing 50 laparoscopic cholecystectomy videos with frame-level annotations for 100 distinct triplet classes. Following the standard evaluation protocol [12], we use videos [2,6,14,23,25,50,51,66,79,111] for testing and the remaining 40 videos for training. The training set contains 78,968 frames while the test set contains 21,895 frames, representing expert-level surgical demonstrations at 1 FPS sampling.

2.3 Dual-Task Autoregressive Imitation Learning (DARIL)

Our IL baseline follows an offline Behavior Cloning (BC) approach, using supervised learning on expert demonstrations to model surgical action prediction as causal sequence generation, combining frame-level processing with autoregressive action generation:

$$p(a_{t+1}|f_{t-w+1:t}) = \text{GPT-2}(\text{FrameEmb}(f_{t-w+1:t})) \tag{1}$$

where $w = 20$ represents the context window size.

Architecture: The model processes 1024-dimensional Swin transformer features [8] through: (1) BiLSTM encoder for temporal current action recognition, (2) GPT-2 decoder [14] for causal future action generation using a context window of $w = 20$ frame embeddings as input, and (3) separate prediction heads for the combined <instrument, verb, and target> class and surgical phase components.

Training: Dual-task optimization combines current action recognition and next action prediction losses, with additional auxiliary losses:

$$\mathcal{L} = \mathcal{L}_{\text{current}} + \mathcal{L}_{\text{next}} + \mathcal{L}_{\text{embed}} + \mathcal{L}_{\text{phase}} \tag{2}$$

where $\mathcal{L}_{\text{current}} = -\sum_t \log p(a_t|f_{t-w+1:t})$ and $\mathcal{L}_{\text{next}} = -\sum_t \log p(a_{t+1}|f_{t-w+1:t})$ represent cross-entropy losses for direct prediction of the 100 action classes, $\mathcal{L}_{\text{embed}} = \sum_t ||e_{t+1} - \hat{e}_{t+1}||^2$ is the MSE loss for next frame embedding prediction, and $\mathcal{L}_{\text{phase}}$ is the phase recognition loss.

2.4 Reinforcement Learning Approaches

For reproducibility, we detail our RL problem formulation: we define states as sequences of frame embeddings, actions as predicted triplets, and design reward functions based on expert demonstration matching using cosine similarity between predicted and ground truth action sequences.

Latent World Model + RL: Following Dreamer [3], the current state-of-the-art for image-based RL tasks, we learn an action-conditioned world model predicting future states and rewards: $p(s_{t+1}, r_t|s_t, a_t)$. PPO [17] trains policies in the learned latent environment with rewards designed for expert demonstration matching.

Direct Video RL: Model-free RL applied directly to video sequences using expert demonstration matching rewards. We evaluate PPO [17] and A2C [9]. These algorithms were selected as representative model-free methods with proven stability for continuous control tasks. We performed careful hyperparameter optimization, treating frame sequences as states and predicted triplets as actions.

Inverse RL Enhancement: Maximum Entropy IRL [20] learns reward functions from expert trajectories. We generate negative examples by sampling actions deviating from expert demonstrations, then use learned rewards to guide policy optimization while maintaining IL baseline performance.

2.5 Evaluation Framework

Recognition Evaluation: Standard mAP computation on current and next action predictions using IVT metrics [12].

Planning Evaluation: Multi-horizon assessment across 1s, 2s, 3s, 5s, 10s, and 20s using mAP degradation analysis.

Component Analysis: Individual performance analysis for instruments, verbs, targets, and their combinations (IV, IT, IVT) to understand degradation patterns.

3 Results

3.1 Main Comparative Results

Table 1 presents our experimental findings. DARIL achieves 34.6% action triplet recognition mAP and 33.6% next frame prediction mAP, consistently outperforming all RL variants across planning horizons. The performance gaps are substantial-world model RL drops to 3.1% at 10s while DARIL maintains 29.2%.

Table 1. IL vs RL Performance Comparison. All values are IVT mAP (%). Current refers to action triplet recognition, 1s/5s/10s refer to next frame prediction at different horizons.

Method	Current	1s	5s	10s
DARIL (Ours)	**34.6**	**33.6**	**31.2**	**29.2**
DARIL + IRL	33.1	32.1	29.6	28.1
DARIL + Direct Video RL	33.2	22.6	19.3	15.9
Latent World Model + RL	33.1	14.0	9.1	3.1

Table 2. DARIL Component-wise Performance Analysis. Current refers to action triplet recognition, Next refers to next frame prediction.

Component	Current	Next	Component	Current	Next
Instrument (I)	91.4	88.2	Instrument-Verb (IV)	42.9	38.8
Verb (V)	69.4	68.1	Instrument-Target (IT)	43.5	43.6
Target (T)	52.7	52.5	IVT	34.6	33.6

3.2 Component-Wise Analysis

Table 2 shows DARIL's component-wise performance. Instruments demonstrate highest stability (91.4% to 88.2%), while targets show more variability (52.7% to 52.5%). The IVT combination reflects multiplicative effects of constituent components.

3.3 Planning Performance Analysis

Figure 1 demonstrates DARIL's smooth degradation across horizons from 33.6% at 1s to 29.2% at 10s (13.1% relative decrease). This demonstrates the model's robust temporal consistency across different planning horizons.

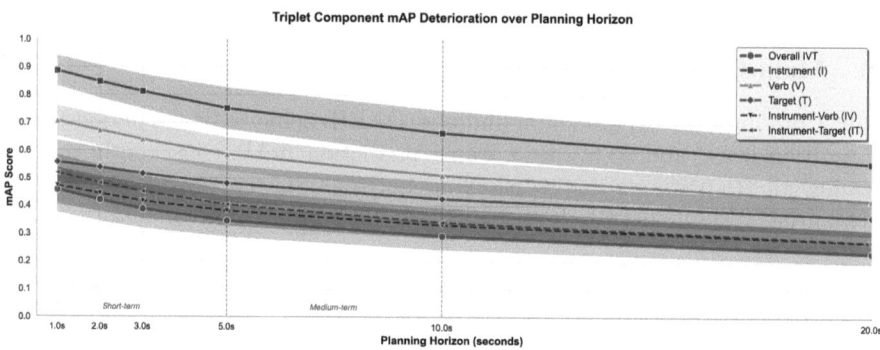

Fig. 1. DARIL planning performance across time horizons. The model maintains stable performance across different IVT mAP score components with graceful degradation over longer planning horizons. Error bars indicate 95% confidence intervals.

3.4 Qualitative Analysis

Figure 2 presents qualitative examples showing DARIL's recognition and planning capabilities. The model correctly identifies current actions while maintaining reasonable planning accuracy for short-term predictions, with graceful degradation over longer horizons.

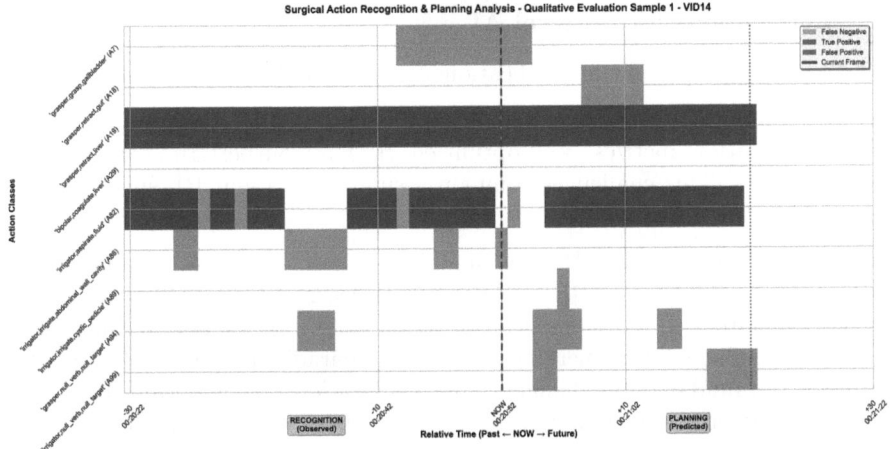

Fig. 2. Qualitative evaluation showing recognition (past) and planning (future) performance. Green indicates true positives, blue shows false positives, beige represents false negatives. The model demonstrates strong current recognition with smooth planning degradation. (Color figure online)

4 Discussion

4.1 Analysis: Why RL Underperformed

Our analysis identifies key factors explaining RL's underperformance:

Expert-Optimal Demonstrations: CholecT50 contains expert-level data already near-optimal for evaluation metrics. RL exploration discovers valid alternatives that appear suboptimal under expert-similarity metrics.

Evaluation Metric Alignment: Test metrics directly reward expert-like behavior, giving IL fundamental advantages. This is common in medical domains where expert behavior defines gold standards.

Limited Exploration Benefits: Surgical domains have strong safety constraints limiting exploration benefits. While RL may discover novel approaches, these appear suboptimal for standard evaluation criteria.

State-Action Representation Challenges: Our RL implementations used frame embeddings as states and action triplets as discrete actions, with reward functions based on expert demonstration similarity. This design faced difficulties with comprehensive state representation and reward signal sparsity, potentially limiting learning effectiveness.

Distribution Mismatch: RL policies trained on different objective functions may produce valid but different behaviors that test metrics penalize due to expert demonstration alignment.

4.2 Implications for Surgical AI

Our findings have significant implications for surgical AI development:

Method Selection: In expert domains with high-quality demonstrations and aligned evaluation metrics, well-optimized IL may outperform sophisticated RL approaches. This challenges common assumptions about RL superiority in sequential decision making.

Bootstrapping RL with IL: Our findings suggest a promising approach for surgical AI: bootstrapping RL models with IL-learned basic skills, then using physics simulators or world models for safe exploration of new techniques, a strategy that aligns with emerging data-centric paradigms in the field [2]. This addresses the fundamental challenge that exploration and mistakes are crucial for improving surgical techniques, but must be tested in simulation rather than on patients.

Safety Considerations: IL approaches inherently stay closer to expert behavior, offering potential safety advantages in clinical deployment. RL exploration introduces uncertainty that may be undesirable in safety-critical surgical contexts.

Clinical Translation: Simpler IL models are easier to validate, interpret, and deploy in clinical settings compared to complex RL systems with learned reward functions.

Cross-Dataset Generalization: RL may prove advantageous when working across multiple datasets due to its ability to generalise more effectively, especially on non-expert trajectories. This suggests potential future work directions where RL's exploration capabilities could be beneficial for handling diverse surgical scenarios and skill levels.

4.3 Limitations and Future Work

Several limitations should be considered: (1) Single dataset evaluation on CholecT50 may not generalise to other surgical procedures. (2) Expert test data similar to training distributions may favor IL—results might differ with sub-expert or out-of-distribution scenarios. (3) Evaluation metrics directly reward expert-like behavior—alternative criteria focusing on patient outcomes might favor RL. (4) More sophisticated RL implementations with better state representations and reward design might outperform IL. (5) Overfitting concerns: Our models likely overfit to the limited number of videos and may not generalise well to test videos, indicating need for larger datasets and better simulators, whether world models or physics engines. (6) Lack of reward feedback: Our RL approaches suffered from insufficient reward signals from possible future states and lack of outcome data, limiting their learning effectiveness.

Future work should explore diverse surgical datasets, develop outcome-focused evaluation metrics, and investigate advanced RL techniques specifically designed for expert domains with comprehensive state-action-reward modeling.

5 Conclusion

This work provides crucial insights for surgical AI by demonstrating conditions where sophisticated RL approaches do not universally improve upon well-optimized IL. Our DARIL baseline consistently outperforms RL variants across planning horizons, with world model RL showing particularly poor performance (3.1% vs 29.2% at 10s).

The key insight is that expert domains with high-quality demonstrations may not benefit from RL exploration when evaluation metrics reward expert-like behavior. Distribution matching on expert-annotated test sets systematically favors IL over potentially valid RL policies that differ from training demonstrations.

Future surgical AI development should carefully consider domain characteristics, data quality, and evaluation alignment when choosing between IL and RL approaches. While IL excels at expert behavior cloning, RL's exploration capabilities may prove valuable in comprehensive evaluation frameworks capturing patient outcomes beyond expert similarity.

References

1. Boels, M., Liu, Y., Dasgupta, P., Granados, A., Ourselin, S.: Swag: long-term surgical workflow prediction with generative-based anticipation. Inter. J. Comput. Assisted Radiol. Surgery, 1–11 (2025)
2. Boels, M., Robertshaw, H., Booth, T.C., Granados, A., Dasgupta, P., Ourselin, S.: Surgical robot learning: From demonstration and simulation to world models-a review. arXiv preprint (2025)
3. Hafner, D., Pasukonis, J., Ba, J., Lillicrap, T.: Mastering diverse domains through world models. arXiv preprint arXiv:2301.04104 (2023)
4. Hansen, N., Su, H., Wang, X.: Td-mpc2: scalable, robust world models for continuous control. arXiv preprint arXiv:2310.16828 (2023)
5. Hu, A., et al.: Gaia-1: A generative world model for autonomous driving. arXiv preprint arXiv:2309.17080 (2023)
6. Kiyasseh, D., et al.: A vision transformer for decoding surgeon activity from surgical videos. Nat. Biomed. Eng. **7**(6), 780–796 (2023)
7. Liu, Y., et al.: Skit: a fast key information video transformer for online surgical phase recognition. In: Proceedings of the IEEE/CVF International Conference on Computer Vision, pp. 21074–21084 (2023)
8. Liu, Z., et al.: Swin transformer: hierarchical vision transformer using shifted windows. In: Proceedings of the IEEE/CVF International Conference on Computer Vision, pp. 10012–10022 (2021)
9. Mnih, V., et al.: Asynchronous methods for deep reinforcement learning. In: International Conference on Machine Learning, pp. 1928–1937. PmLR (2016)
10. Nwoye, C.I., et al.: Cholectriplet 2021: a benchmark challenge for surgical action triplet recognition. Med. Image Anal. **86**, 102803 (2023)
11. Nwoye, C.I., et al.: Recognition of instrument-tissue interactions in endoscopic videos via action triplets. In: Martel, A.L., et al. (eds.) MICCAI 2020. LNCS, vol. 12263, pp. 364–374. Springer, Cham (2020). https://doi.org/10.1007/978-3-030-59716-0_35

12. Nwoye, C.I., Padoy, N.: Data splits and metrics for method benchmarking on surgical action triplet datasets. arXiv preprint arXiv:2204.05235 (2022)
13. Nwoye, C.I., et al.: Rendezvous: attention mechanisms for the recognition of surgical action triplets in endoscopic videos. Med. Image Anal. **78**, 102433 (2022)
14. Radford, A., Wu, J., Child, R., Luan, D., Amodei, D., Sutskever, I., et al.: Language models are unsupervised multitask learners. OpenAI blog **1**(8), 9 (2019)
15. Robertshaw, H., et al.: Reinforcement learning for safe autonomous two-device navigation of cerebral vessels in mechanical thrombectomy. Inter. J. Comput. Assisted Radiol. Surgery, 1–10 (2025)
16. Robertshaw, H., Karstensen, L., Jackson, B., Granados, A., Booth, T.C.: Autonomous navigation of catheters and guidewires in mechanical thrombectomy using inverse reinforcement learning. Int. J. Comput. Assist. Radiol. Surg. **19**(8), 1569–1578 (2024)
17. Schulman, J., Wolski, F., Dhariwal, P., Radford, A., Klimov, O.: Proximal policy optimization algorithms. arXiv preprint arXiv:1707.06347 (2017)
18. Shi, C., Zheng, Y., Fey, A.M.: Recognition and prediction of surgical gestures and trajectories using transformer models in robot-assisted surgery. In: 2022 IEEE/RSJ International Conference on Intelligent Robots and Systems (IROS), pp. 8017–8024. IEEE (2022)
19. Weerasinghe, K., Roodabeh, S.H.R., Hutchinson, K., Alemzadeh, H.: Multimodal transformers for real-time surgical activity prediction. In: 2024 IEEE International Conference on Robotics and Automation (ICRA), pp. 13323–13330. IEEE (2024)
20. Ziebart, B.D., Maas, A.L., Bagnell, J.A., Dey, A.K., et al.: Maximum entropy inverse reinforcement learning. In: AAAI, Chicago, IL, USA, vol. 8, pp. 1433–1438 (2008)

Temporal Propagation of Asymmetric Feature Pyramid for Surgical Scene Segmentation

Cheng Yuan and Yutong Ban(✉)

Global College, Shanghai Jiao Tong University, Shanghai, China
yban@sjtu.edu.cn

Abstract. Surgical scene segmentation is crucial for robot-assisted laparoscopic surgery understanding. Current approaches face two challenges: (i) static image limitations including ambiguous local feature similarities and fine-grained structural details, and (ii) dynamic video complexities arising from rapid instrument motion and persistent visual occlusions. While existing methods mainly focus on spatial feature extraction, they fundamentally overlook temporal dependencies in surgical video streams. To address this, we present temporal asymmetric feature propagation network, a bidirectional attention architecture enabling cross-frame feature propagation. The proposed method contains a temporal query propagator that integrates multi-directional consistency constraints to enhance frame-specific feature representation, and an aggregated asymmetric feature pyramid module that preserves discriminative features for anatomical structures and surgical instruments. Our framework uniquely enables both temporal guidance and contextual reasoning for surgical scene understanding. Comprehensive evaluations on two public benchmarks show the proposed method outperforms the current SOTA methods by a large margin, with +16.4% mIoU on EndoVis2018 and +3.3% mAP on Endoscapes2023. Code is available at https://github.com/cyuan-sjtu/ViT-asym.

Keywords: Surgical scene segmentation · Bidirectional attention · Temporal feature propagation

1 Introduction

The advancement of robot-assisted minimally invasive laparoscopic surgery has heightened the importance of precise surgical scene segmentation, which serves as a fundamental prerequisite for subsequent understanding tasks including pose estimation [3], triplet recognition [7], and safety assessment [1]. In addition, separately for augmented reality and automatic annotation, referring to the scene segmentation mask can reduce render errors and labor consumption. However, achieving accurate scene segmentation is challenging, due to the local feature similarity among various anatomies and the fine-grained structure complexity of

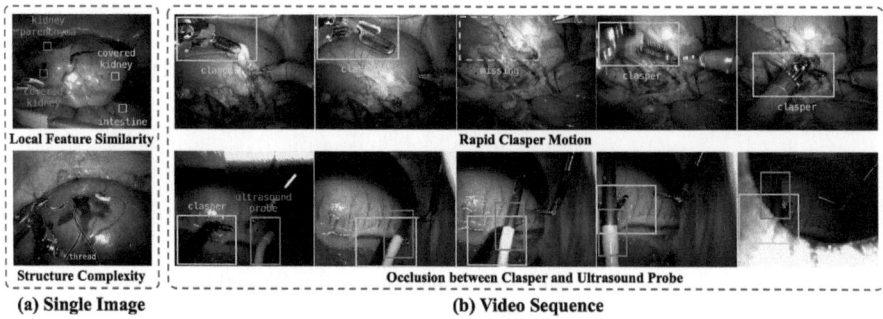

Fig. 1. Challenges in surgical scene segmentation: (a) local feature similarity and fine-grained structure complexity in a single image; (b) rapid object motion and inevitable interaction occlusion in video sequences.

deformable instruments, as shown in Fig. 1(a). Moreover, the instance recognition performance further degrades because of blur from rapid instrument manipulation and inevitable interaction occlusion, as shown in Fig. 1(b).

To overcome these challenges, existing methods have evolved along two main methodological paradigms: (1) The first category employs specialized convolution networks that maximize spatial information modeling through hierarchical feature extraction. Yang et al. [24] implemented atrous spatial pyramid to capture multiscale context representations. Ni et al. [17] proposed a space-squeeze reasoning network incorporating low-rank bilinear feature fusion for different surgical region segmentation. Recent innovation by Liu et al. [15] introduced a large kernel attention strategy to long-strip targets, establishing superior results in instrument-tissue segmentation. (2) The second category capitalizes the transformer architecture, adapting its attention mechanism to get global information perception in segmentation. Groundbreaking work by Carion et al. [5] established an end-to-end query-based detection framework through bipartite matching loss. Pioneering segmentation methods like Mask2Former [6] and MaskDINO [14] demonstrated the effectiveness and versatility of query-based dense predictions. These methods achieved great performance in the natural image field, while they generally disable when meeting specific challenges of surgical videos.

Surgical videos contain more plentiful information compared to a single static image, especially temporal clues. Existing methods proved the effectiveness of temporal consistency in several surgical vision tasks, such as phase recognition [12,20] and instrument tracking [11,19]. However, only a limited number of temporal networks on pixel-level dense prediction appeared recently. Jin et al. [13] extended the transformer architecture through intra- and inter-video relation modeling to capture global dependencies. It focused on tackling the class imbalance problem and produced relatively coarse segmentation. Thus, how to construct temporal propagation and incorporate it into feature enhancement is essential to achieve superior segmentation performance.

We present Temporal Asymmetric Feature Propagation Network (TAFPNet) for surgical video segmentation, addressing spatial ambiguity between anatomy and instruments. It incorporates two perception enhancement modules into a bidirectional attention network architecture. Temporal coherence is maintained via dynamic query propagation, forming occlusion-resistant feature tubes. This integration of structural asymmetry and temporal continuity enables accurate segmentation of both anatomical tissue and dynamic instruments in challenging laparoscopic scenarios. Our main contributions are summarized as follows:

1. We propose TAFPNet, a novel surgical scene segmentation framework that combines temporal dynamics with structural asymmetry.
2. The designed temporal query propagator integrated temporal dependency into the transformer encoding. Meanwhile, the aggregated asymmetric feature pyramid integrates the disentangled instrument-tissue features along the time dimension.
3. The proposed TAFPNet achieves the best overall performance by +16.4% mIoU on EndoVis2018 [2] and +3.3% mAP on Endoscapes2023 [16]. It outperforms in granular structure segmentation with 14.8% and 4.5% of improvement on hepatocystic triangle and cystic artery, respectively.

2 Method

In this section, we present the proposed TAFPNet, which incorporates the Temporal Query Propagator (TQP) and the Aggregated Asymmetric Feature Pyramid (AAFP) to a bidirectional attention network for accurate surgical scene segmentation from laparoscopic surgery videos. The detailed architecture of TAFPNet is illustrated in Fig. 2.

2.1 Bidirectional Attention Architecture

We construct a bidirectional attention architecture to take advantage of progressive feature interaction between a dual-feature extraction branch, a transformer encoding, and a convolution pyramid, to a large extent. Given an image sequence, the multiscale feature pyramid is generated by a stack of 3D convolutions at resolutions of $\{1/4, 1/8, 1/16, 1/32\}$ relative to the input. Subsequently, a multistage feature interaction process is constructed to fuse multi-perceptive information, where each stage comprises two pathways. In the transformer encoding, the input feature embedding passes through the TQP module, multihead attention, and dimensional expansion, then added to the output of the convolution branch. Meanwhile, in the convolution encoding, the input feature pyramid is enhanced by the AAFP module and then compressed into the same dimension embedding to add to the output of transformer branch. After M-stage bidirectional attention, fused features are fed into the transformer decoder for mask prediction. Therefore, it enables the progressive refinement of spatial precision and temporal coherence in surgical video, by implementing parameter-shared residual connection and element-wise summation.

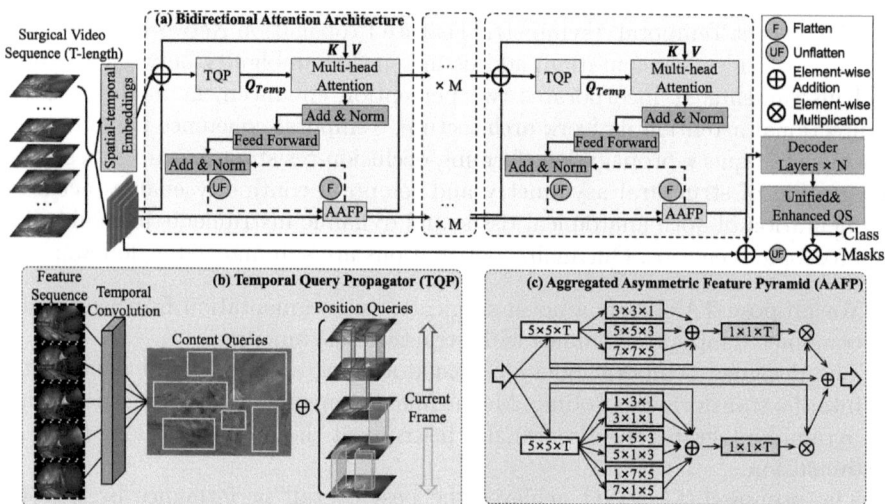

Fig. 2. The overall framework of TAFPNet. It contains a (a) bidirectional attention architecture injected with the (b) Temporal Query Propagator (TQP) and the (c) Aggregated Asymmetric Feature Pyramid (AAFP) module.

2.2 Temporal Query Propagator

The way of incorporating temporal coherence in surgical video analysis is inherently crucial to leverage its benefits. For such purposes, we design a temporal query propagator, namely TQP, which introduces temporal information into the query generation and propagation. This strategy ensures consistent segmentation results across observed frames, so it can generalize well to some cases of rapid instrument manipulation and anatomy-instrument occlusion.

Given a sequence of T length consecutive frames, we first extract frame-wise features F_1, F_2, F_T by a ResNet-50 backbone. The obtained features are then projected with learnable weights W^K and W^V. TQP is an attention module which is further applied to propagate the projected feature along the temporal axis. The key K and the value V of the TQP are calculated based on feature pyramid concat$\{F_t\}_{t=1}^{T}$. The temporal query Q is a learned embedding, which is designed to efficiently aggregate the information in regions of interest along the temporal axis. It contains two parts, the content query Q^{cont} and the position query Q^{pos}. The content query is calculated by:

$$Q^{content} = \text{top-}K(\text{Conv}\left(\text{concat}\{F_t\}_{t=1}^{T}\right)). \tag{1}$$

where top-$K(\cdot)$ represents top-k query selection based on the high activation score, and concat$(\cdot)$ represents concatenation along with time dimension. The position query of the t-th frame represents the position embedding of a spatial-temporal tube, which is obtained by:

$$Q_t^{pos} = \begin{cases} \text{Linear}\left(Q_{t+1}^{pos}\right) & \text{if } 1 \leq t \leq (T-1)/2 \\ W^Q \text{Conv}\left(\text{concat}\{F_t\}_{t=1}^T\right) & \text{if } t = (T+1)/2 \\ \text{Linear}\left(Q_{t-1}^{pos}\right) & \text{if } (T+1)/2 < t \leq T \end{cases} \quad (2)$$

where W^Q is a learnable weight. Subsequently, we feed the content query, the position query, the key, and the value into multihead attention layers for feature refinement:

$$\text{Attention}\left(Q, K, V\right) = \text{Softmax}\left(\frac{(\sum_{t=1}^T Q_t^{pos}) \odot Q^{content} K^T}{\sqrt{d_k}}\right) V \quad (3)$$

where $\odot$ represents the dot product operator. After training, each of the selected temporal queries corresponds to a spatial-temporal feature tube, see Fig. 2 (b).

2.3 Aggregated Asymmetric Feature Pyramid

Characteristics of anatomy and instrument generally have obvious differences in visual perspective, presenting irregular-polygon and regular-bar shape, separately [25]. In addition, the same anatomy or instrument maintains inherent topological structure invariance in a video sequence. Thus, we design an aggregated asymmetric Feature Pyramid, namely AAFP, to comprehensively aggregate frame features, from global to local. It contains two perception enhancement operations for anatomy and instruments, separately.

The input feature pyramid concat$\{F_t\}_{t=1}^T$ first passes through a convolution with the kernel size of $(5 \times 5 \times T)$ to obtain a spatial-temporal aggregating feature map F_{temp}. For anatomy perception, we adopt symmetric convolutions to enhance irregular-polygon features:

$$F_m^{AP} = \text{Conv}_{k_m \times k_m \times t_m}\left(F_{temp}\right) \quad (4)$$

where multiple (k_m, k_m, t_m) pairs are set as $(3,3,1)$, $(5,5,3)$, and $(7,7,5)$. For instrument perception, we adopt asymmetric calculations by parallel strip convolution pairs to enhance the regular-bar features:

$$F_m^{IP} = \text{Conv}_{k_m \times 1 \times t_m}\left(F_{temp}\right) + \text{Conv}_{1 \times k_m \times t_m}\left(F_{temp}\right) \quad (5)$$

where k_m and t_m are same with them in symmetric convolutions. Subsequently, the final aggregated attention map of registered as E_{temp} is calculated by:

$$E_{temp} = \text{Conv}_{1 \times 1 \times T}\left(\sum_{m=0}^M \left(F_m^P\right) + F_{temp}\right) \bigotimes F_{temp} \quad (6)$$

where M, $\bigotimes$, and F_m^P represents the number of selected kernels, matrix multiplication, and the enhanced feature map of anatomy or instruments. Finally, aggregated attention maps from both anatomy and instrument perception blocks are added as the final output.

3 Experiments and Results

Datasets and Evaluation Metrics. The experiments were performed on two laparoscopic surgery datasets, and our proposed method was conducted a comparative analysis with existing state-of-the-art (SOTA) methods.

The first dataset named EndoVis2018 [2] consists of 19 sequences, officially divided into 15 training sets and 4 test sets. Each sequence contains about 250 stereo-pair frames with 1280 × 1024 resolution. The segmentation ground truth is released by the challenge organizer, and here we only utilize the annotated left image sequences. This segmentation task is defined to divide the entire surgical scene into 12 categories, including different anatomies and instruments.

The second dataset named Endoscapes2023 [16] was recently released for three sub-tasks, that are surgical scene segmentation, object detection, and critical view of safety assessment. Here, we use its subset consisting of 493 annotated frames from 50 laparoscopic cholecystectomy videos. It is officially divided into 3:1:1 ratio as the training set, validation set, and test set. Compared to EndoVis2018, this segmentation task is more challenging as it requires the identification of six distinct types of anatomical structures and instruments from the entire image.

To capture temporal clues, we utilize the temporal window to cover 5% of the mean sequence duration as offline datasets. In the test, we follow the same evaluation manner as the challenge required for a direct and fair comparison. We also use the official evaluation protocol, i.e., (1) mean intersection-over-union (mIoU) and mean Dice coefficient (mDice) for EndoVis2018, and (2) both detection and segmentation mean average precision (mAP@[0.5:0.95]) for Endoscapes2023.

Implementation Details. The resolution of EndoVis2018 and Endoscapes2023 image is reduced to 512 × 1024 resolution for memory saving. Our proposed method is implemented in PyTorch (2.2.1 version) with 2 NVIDIA A40 GPUs for calculation. Double GPUs only enable the network to be trained in the batch size of 4. Model training iteration and base learning rate are set to 20k and 1e-4, separately. Deeper networks did not produce obvious improvements in early trials, so the backbone of our network is stuck to ResNet-50 [10].

Comparison with State-of-the-Art Methods. We compare our TAFP-Net with SOTA methods. On EndoVis2018, Table 1 lists the compared methods in the first line, involving in three categories: (1) Three reported methods in the 2018 Robot Scene Segmentation Challenge like NCT, UNC, and OTH [2]; (2) Multiscale feature fusion networks such as U-Net [18], DeepLabv3+ [24], UPerNet [22], and HRNet [21]; (3) Specific-designed segmentation methods, such as SegFormer [23], SegNeXt [8], STswinCL [13], and LSKANet [15]. Our TAFP-Net consistently outperforms in all four test sequences and exceeds the previous SOTA results by 16.4% at mIoU and 14.6% at mDice. In particular, the mIoU of sequence 2 and 4 outperforms the previous best performance with a huge improvement of 21.6% and 31.7%. Although compared methods constructed advanced spatial feature enhancement strategies, our TAFPNet combines temporal dynamics with structural asymmetry to explicitly generate reliable guidance for segmentation.

Table 1. Quantitative results on EndoVis2018. The best results are bold, and the second best results are underlined.

Method		OTH [2]	U-Net [18]	DeepLab v3+ [24]	UPer Net [22]	HR Net [21]	SegFormer [23]	SegNe Xt [8]	STswin CL [13]	LSKA Net [15]	Base Net	AFP Net	TAFPNet (Ours)
mIoU (%)	Seq1	69.1	55.6	64.1	67.4	68.9	69.1	<u>70.6</u>	67.0	67.5	54.1	70.5	**76.2**
	Seq2	57.5	50.5	57.0	54.0	57.3	55.8	57.1	63.4	63.2	63.8	<u>83.1</u>	**85.4**
	Seq3	82.9	69.7	82.3	82.0	85.0	81.6	84.3	83.7	<u>85.2</u>	70.8	83.8	**88.1**
	Seq4	39.0	26.8	31.6	30.2	42.1	45.5	45.1	40.3	48.9	47.0	<u>72.4</u>	**80.6**
	Overall	62.1	50.7	58.8	58.4	63.3	63.0	64.3	63.6	66.2	58.9	<u>77.5</u>	**82.6**
mDice (%)		-	61.5	67.3	66.8	71.8	71.9	72.5	72.0	75.3	69.3	<u>86.6</u>	**89.9**

Table 2 shows the performance of compared methods on Endoscapes2023. The compared methods consist of three fine-tuned segmentation models, namely Mask-RCNN [9], Cascade Mask-RCNN [4], and Mask2Former [6]. Although the detection and segmentation performance of each category differs greatly in the compared methods, our TAFPNet achieves the best overall results with the detection mAP of 30.8% and the segmentation mAP of 29.9%. However, a substantial performance gap persists in fine-grained object recognition, revealing the extra difficulties of Endoscapes2023, probably due to worse lighting situation and more motion blurs.

Table 2. Quantitative results on Endoscapes2023. The best results are bold, and the second best results are underlined.

Method	Detection mAP@[0.5:0.95] | Segmentation mAP@[0.5:0.95] (%)						
	Cystic Plate	HC Triangle	Cystic Artery	Cystic Duct	Gallbladder	Tool	Overall
Mask-RCNN [9]	2.8 | 3.3	2.9 | 3.8	12.7 | 11.9	7.4 | 7.9	45.8 | 59.1	49.7 | 51.2	20.2 | 22.9
Cascaded Mask-RCNN [4]	<u>3.1</u> | **6.5**	11.1 | 6.7	<u>15.7</u> | 10.8	11.7 | 10.0	62.0 | 62.7	63.2 | 56.8	27.8 | 25.6
Mask2Former [6]	1.4 | 1.7	<u>14.3</u> | 8.3	6.4 | 7.6	<u>14.7</u> | **15.9**	68.7 | 62.6	67.1 | **63.5**	28.8 | 26.6
BaseNet	1.1 | 0.5	12.8 | 10.4	9.8 | 12.2	7.2 | 7.9	64.0 | <u>63.2</u>	58.7 | 59.5	25.6 | 25.6
AFPNet	2.8 | 3.0	5.6 | <u>10.7</u>	**16.5** | **20.1**	**14.9** | 10.1	<u>67.1</u> | **65.4**	**72.2** | 62.2	<u>29.8</u> | <u>28.7</u>
TAFPNet (Ours)	**10.5** | <u>6.3</u>	**16.3** | **23.1**	14.9 | <u>16.4</u>	10.6 | <u>10.3</u>	63.5 | 60.5	<u>68.9</u> | <u>62.8</u>	**30.8** | **29.9**

Effectiveness of Temporal Propagation and Asymmetric Attention. We evaluate the effectiveness of two core enhancement modules in our TAFPNet. The results of three ablation settings are also listed in Table 1 and Table 2: (1) a plain vision transformer encoder-decoder network as the baseline (BaseNet), (2) an Asymmetric Feature Propagation Network (AFPNet) without temporal coherence guidance, and (3) our proposed framework (TAFPNet). We adopt the same backbone for different settings for fairness. It can be seen that AFPNet performs better than BaseNet, especially on EndoVis2018, demonstrating the effectiveness of the bidirectional attention architecture. Furthermore, our proposed TAFPNet further increases performance in almost all categories, except for the cystic artery and gallbladder in Endoscapes2023. Although dim ambient light causes low recognition precision scores on Endoscapes2023, our proposed TAFPNet still identifies delicate structures such as the cystic artery and

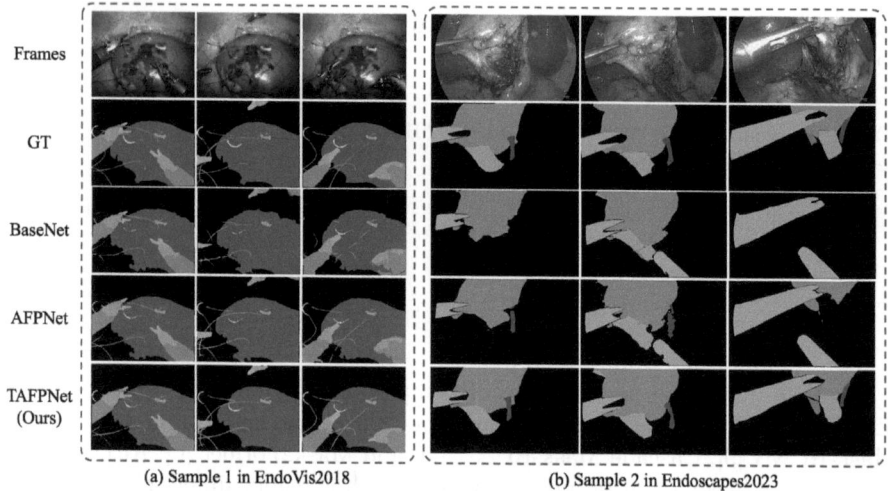

Fig. 3. Visual comparison of segmentation results on (a) EndoVis2018 and (b) Endoscapes2023. From top to bottom, for each dataset, we present three continuous video frames and their corresponding ground truth, with segmentation results using BaseNet, AFPNet and our proposed TAFPNet.

the cystic plate, which are terribly missed in compared ablation networks. This demonstrates that temporal propagation significantly enhances feature propagating precision. Figure 3 exhibits some visual results of ablation experiments. TAFPNet can achieve the complete and consecutive segmentation of slender thread in sequence 1, which is misidentified in other ablation networks. Besides, the blurred visual boundary of rapidly moved clasper in sequence 2 is more accurately and clearly recognized by our TAFPNet. In summary, our proposed TAFPNet shows a noticeable advancement in the ability of fine-grained structure recognition and fast motion perception.

4 Conclusion

In this work, we propose a novel interaction framework for surgical scene segmentation that incorporates temporal dependency in query propagation and disentangled feature representation. Our method achieves the best overall performance in both scene segmentation on EndoVis2018 and object recognition on Endoscapes2023, especially on fine-grained structures. The superior performance establishes a promising value for clinical robot-assisted intervention.

References

1. Ali, S., Zhou, F., Bailey, A., et al.: A deep learning framework for quality assessment and restoration in video endoscopy. Med. Image Anal. **68**, 101900 (2021)

2. Allan, M., Kondo, S., Bodenstedt, S., et al.: 2018 robotic scene segmentation challenge. arXiv preprint arXiv:2001.11190 (2020)
3. Allan, M., Ourselin, S., Hawkes, D.J., et al.: 3-D pose estimation of articulated instruments in robotic minimally invasive surgery. IEEE Trans. Med. Imaging **37**(5), 1204–1213 (2018)
4. Cai, Z., Vasconcelos, N.: Cascade R-CNN: high quality object detection and instance segmentation. IEEE Trans. Pattern Anal. Mach. Intell. **43**(5), 1483–1498 (2019)
5. Carion, N., Massa, F., Synnaeve, G., Usunier, N., Kirillov, A., Zagoruyko, S.: End-to-end object detection with transformers. In: Vedaldi, A., Bischof, H., Brox, T., Frahm, J.-M. (eds.) ECCV 2020. LNCS, vol. 12346, pp. 213–229. Springer, Cham (2020). https://doi.org/10.1007/978-3-030-58452-8_13
6. Cheng, B., Schwing, A., Kirillov, A.: Per-pixel classification is not all you need for semantic segmentation. Adv. Neural. Inf. Process. Syst. **34**, 17864–17875 (2021)
7. Collins, T., Pizarro, D., Gasparini, S., et al.: Augmented reality guided laparoscopic surgery of the uterus. IEEE Trans. Med. Imaging **40**(1), 371–380 (2020)
8. Guo, M.H., Lu, C.Z., Hou, Q., et al.: Segnext: rethinking convolutional attention design for semantic segmentation. Adv. Neural. Inf. Process. Syst. **35**, 1140–1156 (2022)
9. He, K., Gkioxari, G., Dollár, P., et al.: Mask r-cnn. In: Proceedings of the IEEE International Conference on Computer Vision, pp. 2961-2969 (2017)
10. He, K., Zhang, X., Ren, S., et al.: Deep residual learning for image recognition. In: IEEE Conference on Computer Vision and Pattern Recognition, pp. 770–778. IEEE (2016)
11. Jin, Y., Cheng, K., Dou, Q., Heng, P.-A.: Incorporating temporal prior from motion flow for instrument segmentation in minimally invasive surgery video. In: Shen, D., et al. (eds.) MICCAI 2019. LNCS, vol. 11768, pp. 440–448. Springer, Cham (2019). https://doi.org/10.1007/978-3-030-32254-0_49
12. Jin, Y., Dou, Q., Chen, H., et al.: SV-RCNet: workflow recognition from surgical videos using recurrent convolutional network. IEEE Trans. Med. Imaging **37**(5), 1114–1126 (2018)
13. Jin, Y., Yu, Y., Chen, C., et al.: Exploring intra-and inter-video relation for surgical semantic scene segmentation. IEEE Trans. Med. Imaging **41**(11), 2991–3002 (2022)
14. Li, F., Zhang, H., Xu, H., Liu, S., et al.: Mask dino: towards a unified transformer based framework for object detection and segmentation. In: Proceedings of the IEEE/CVF Conference on Computer Vision and Pattern Recognition, pp. 3041–3050. IEEE/CVF (2023)
15. Liu, M., Han, Y., Wang, J., et al.: LSKANet: long strip kernel attention network for robotic surgical scene segmentation. IEEE Trans. Med. Imaging **43**(4), 1308–1322 (2023)
16. Murali, A., Alapatt, D., Mascagni, P., et al.: The endoscapes dataset for surgical scene segmentation, object detection, and critical view of safety assessment: official splits and benchmark. arXiv preprint arXiv:2312.12429 (2023)
17. Ni, Z.L., Bian, G.B., Li, Z., et al.: Space squeeze reasoning and low-rank bilinear feature fusion for surgical image segmentation. IEEE J. Biomed. Health Inform. **26**(7), 3209–3217 (2022)
18. Ronneberger, O., Fischer, P., Brox, T.: U-Net: convolutional networks for biomedical image segmentation. In: Navab, N., Hornegger, J., Wells, W.M., Frangi, A.F. (eds.) MICCAI 2015. LNCS, vol. 9351, pp. 234–241. Springer, Cham (2015). https://doi.org/10.1007/978-3-319-24574-4_28

19. Sarikaya, D., Corso, J.J., Guru, K.A.: Detection and localization of robotic tools in robot-assisted surgery videos using deep neural networks for region proposal and detection. IEEE Trans. Med. Imaging **36**(7), 1542–1549 (2017)
20. Twinanda, A.P., Shehata, S., Mutter, D., et al.: Endonet: a deep architecture for recognition tasks on laparoscopic videos. IEEE Trans. Med. Imaging **36**(1), 86–97 (2017)
21. Wang, J., Sun, K., Cheng, T., et al.: Deep high-resolution representation learning for visual recognition. IEEE Trans. Pattern Anal. Mach. Intell. **43**(10), 3349–3364 (2020)
22. Xiao, T., Liu, Y., Zhou, B., et al.: Unified perceptual parsing for scene understanding. In: Proceedings of the European Conference on Computer Vision (ECCV), pp. 418–434 (2018)
23. Xie, E., Wang, W., Yu, Z., et al.: SegFormer: simple and efficient design for semantic segmentation with transformers. Adv. Neural. Inf. Process. Syst. **34**, 12077–12090 (2021)
24. Yang, Z., Peng, X., Yin, Z.: Deeplab v3 plus-net for image semantic segmentation with channel compression. In: IEEE 20th International Conference on Communication Technology (ICCT), pp. 1320-1324. IEEE (2020)
25. Yuan, C., Ban, Y.: Surgical Scene Segmentation by Transformer With Asymmetric Feature Enhancement. In: IEEE 22nd International Symposium on Biomedical Imaging. IEEE (2024)

Author Index

A
Ahn, Jeong Woo 168
Asad, Muhammad 74
Assis, Tiago 105
Ayache, Nicholas 137

B
Ban, Yutong 187
Barrag, Juan Antonio 21
Bastian, Lennart 53
Berthet-Rayne, Pierre 137
Billot, Benjamin 137
Boels, Maxence 178
Booth, Thomas C. 178
Brahaj, Luana 1
Buch, Vivek P. 168
Budd, Charlie 148
Byrd, Grayson 21

C
Cattin, Philippe C. 1
Cha, Richard Jaepyeong 85
Chen, Juo Tung 85
China, D. 158
Cho, Hyun Jin 64
Choi, Juseung 95
Colleoni, Emanuele 74

D
Dasgupta, Prokar 178
De Neve, Wesley 95
Delingette, Hervé 137
Ding, Hao 21
Dorent, Reuben 105
Dou, Qi 11
Durrer, Alicia 1

E
El Hadramy, Sidaty 1
Elliot, Matthew 148
Elson, Daniel S. 116

F
Facente, Federica 137
Farshad, Azade 32
Friedrich, Paul 1

G
Garcia, Nuno C. 105
Goldenberg, Antony 85
Golland, Polina 137
Gopalakrishnan, Vivek 137
Granados, Alejandro 178
Guo, Fengyue 11

H
Hager, Gregory 42
Hasler, Carol C. 1
He, Ziling 11
Heiliger, Christian 53
Heng, Pheng-Ann 126
Hu, Xiaowei 126
Hu, Yicheng 116
Huang, Baoru 116

I
Im, Jiwon 95
Ishii, Masaru 42
Iyer, N. 158

J
Jang, Woowon 95
Jin, Pengfei 64

K
Kattel, Manasi 137
Kazanzides, Peter 21

Kim, G. 158
Kim, Ji Woong Brian 85
Kim, Sekeun 64
Kim, Yoseph 85
Köksal, Çağhan 32
Krieger, Axel 85
Kwon, Jong Yun 168

L
Lee, Hwanhee 168
Lee, J. 158
Lee, Jinhee 168
Lee, Sanghoon 168
Lewandowski, Julia 168
Li, Chengkun 11
Li, Edwin 1
Li, Jinpeng 126
Li, Quanzheng 64
Liu, Lihao 126
Long, Yonghao 11
Luengo, Imanol 74

M
Machado, Ines P. 105
Mangulabnan, Jan Emily 42
Mazomenos, Evangelos B. 74
McGovern, R. 158
McMahon, Ciara 168

N
Navab, Nassir 32, 53

O
Oh, Yujin 64
Ourselin, Sebastien 178
Ozbulak, Utku 95

P
Park, Jay J. 168
Park, Seohee 168
Pei, Jialun 11
Peng, Bin 11

R
Rashidian, Niki 95
Robertshaw, Harry 178

S
Sanchez-Matilla, Ricardo 74
Seenivasan, Lalithkumar 21, 42

Ségaud, Silvère 148
Shapey, Jonathan 148
Shu, Hongchao 21
Soberanis-Mukul, Roger D. 42
Song, Sifan 64
Stasiuk, Graeme 148
Stoyanov, Danail 74
Su, Jionglong 116

T
Taylor, Russell H. 21, 42
Tivnan, Matthew 64
Toussaint, Nicolas 74
Tsai, De Ru 85

U
Unberath, Mathias 21, 42
Uneri, A. 158
Uppot, Raul 64

V
Vedula, S. Swaroop 42
Vercauteren, Tom 148

W
Wang, Guangsuo 11
Wang, Tony Danjun 53
Wang, Yihan 126
Wehrli, Michael 1
Wei, Wen 137
Wu, Dufan 64

X
Xiao, Pu 21
Xie, Yijing 148
Xu, Jialang 74
Xu, Mengya 11
Xu, Songyu 116

Y
Yan, Qiao 126
Yeganeh, Yousef 32
Yoon, Siyeop 64
You, Yiyang 85
Yuan, Cheng 187
Yuan, Yuchen 126

Z
Zhang, Han 21
Zwick, Benjamin 105

MIX
Papier aus verantwortungsvollen Quellen
Paper from responsible sources
FSC® C105338

If you have any concerns about our products,
you can contact us on
ProductSafety@springernature.com

In case Publisher is established outside the EU,
the EU authorized representative is:
**Springer Nature Customer Service Center GmbH
Europaplatz 3, 69115 Heidelberg, Germany**

Printed by Libri Plureos GmbH
in Hamburg, Germany